I0803278

IMAGES
of America

Mount Tamalpais and the Marin Municipal Water District

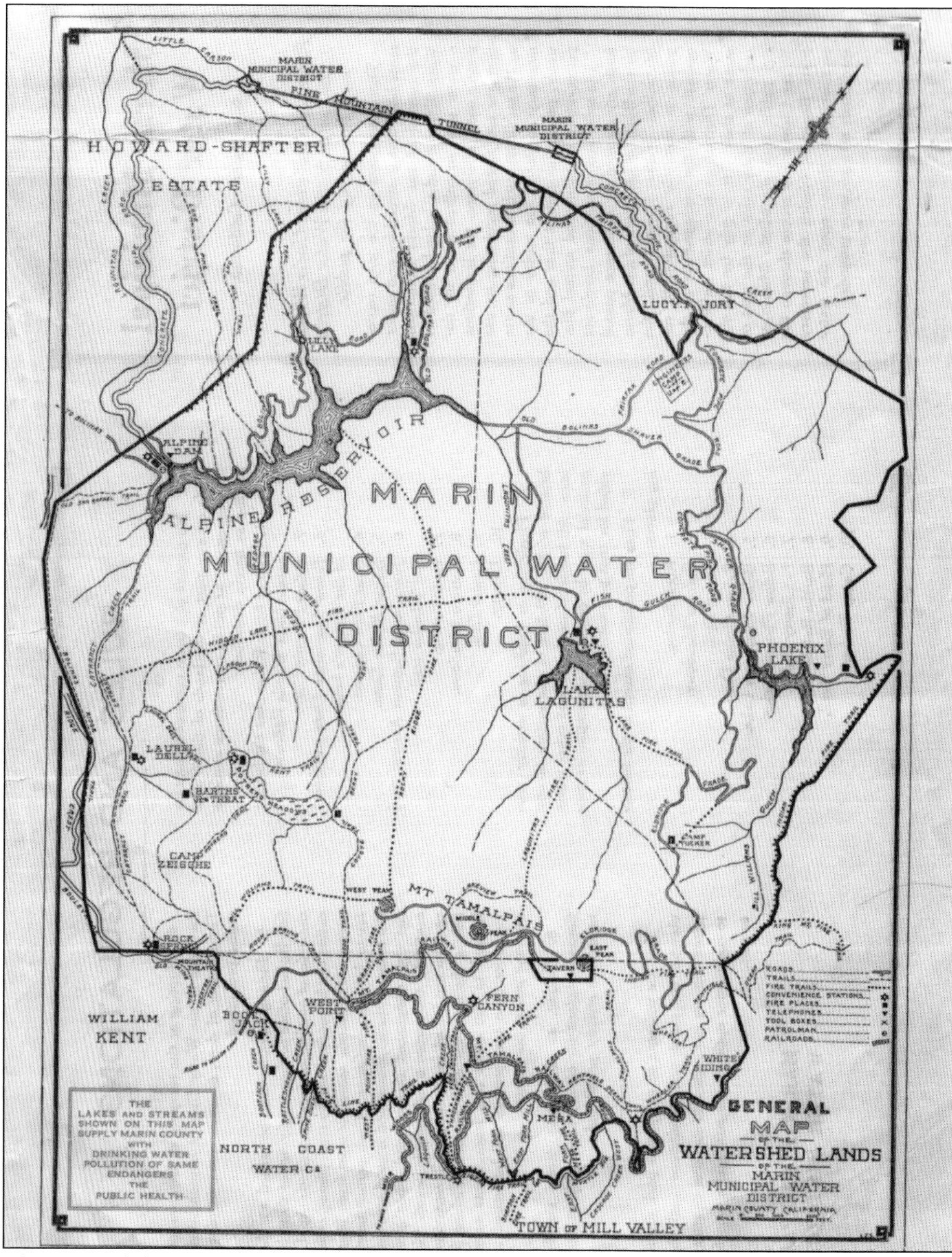

This map of the Marin Municipal Water District watershed was created sometime after 1924. Only three of the district's reservoirs are shown: Lagunitas, Phoenix, and Alpine. Bon Tempe Reservoir, which was built in 1948, is not shown. Note "Engineers Camp," near the upper right-hand portion, which MMWD leased to the University of California Berkeley College of Civil Engineering as a summer school from 1924 to 1948.

ON THE COVER: Alpine Dam, the first of three dams built on Lagunitas Creek by the Marin Municipal Water District, was constructed between the summer of 1917 and the spring of 1919.

IMAGES
of America

MOUNT TAMALPAIS AND THE MARIN MUNICIPAL WATER DISTRICT

Jack Gibson

ARCADIA
PUBLISHING

ISBN 978-0-7385-9317-3

Published by Arcadia Publishing
Charleston, South Carolina

Printed in the United States of America

Library of Congress Control Number: 2011940555

For all general information, please contact Arcadia Publishing:
Telephone 843-853-2070
Fax 843-853-0044
E-mail sales@arcadiapublishing.com
For customer service and orders:
Toll-Free 1-888-313-2665

Visit us on the Internet at www.arcadiapublishing.com

Dedicated to the memory of Robert Walter "Bob" Lethbridge, whose decades of collecting and preserving MMWD history provided much valuable information used in the creation of this book.

Contents

Acknowledgments

Many individuals, libraries, and historical societies were instrumental in the preparation of this book. However, the book would simply not have been possible without the assistance of Elisa Ignatius, Marin Municipal Water District (MMWD) public information representative. It was also facilitated by knowledgeable assistance and encouragement from Ron Theisen, retired MMWD engineering division manager; Libby Pischel, MMWD public information officer; Matt Cerkel, MMWD park ranger; Bill Hogan, MMWD watershed protection manager and chief ranger; Eric McGuire, retired MMWD environmental services coordinator; my water board colleagues; and the many MMWD employees who supported and assisted the project. I would also like to thank MMWD for allowing me to use the photographs in this book and giving me access to its significant repository of historical documents.

I am also grateful to History Room Librarian David Grossman, docent Barbara Ford, and the other staff and docents of the Lucretia Little History Room of the Mill Valley Public Library; Librarian Jocelyn Moss of the Marin History Museum; and Librarian Laurie Thompson and Digital Archivist Carol Acquaviva of the Anne T. Kent California Room, Marin County Free Library, for their help in researching and providing images to tell the story of MMWD. Thank you as well to the Bancroft Library at the University of California, Berkeley, and the National Park Service for contributing several valuable images.

Unless otherwise noted, all images appear courtesy of the Marin Municipal Water District.

INTRODUCTION

Today, the image of Mount Tamalpais serves as the logo for the Marin Municipal Water District. The mountain, the crown jewel of the district's properties, is an icon for both the water district and the county of Marin. The properties owned by the water district—the mountain's watershed and nearby watersheds—total approximately 22,000 acres, covering about seven percent of the land area of Marin County. This land sits adjacent to open space owned by the county, as well as the vast holdings of the Golden Gate National Recreation Area and Mount Tamalpais State Park. Together, they provide a world-class recreation and wilderness area only 30 minutes from downtown San Francisco. In 1912, when the Marin Municipal Water District was chartered, the Tamalpais watershed and the surrounding wilderness areas were available to serve the people of Marin; the land had remained undeveloped throughout the 19th century. Mount Tamalpais, located in the shadow of San Francisco, survived in its natural state throughout the boom of the Gold Rush, the development of San Francisco, and a notorious period of wanton greed for natural resources. The water district's birth in 1912 allowed the continued protection of that wilderness area for the remainder of that century and into the 21st century.

The slow pace of development in Marin was lamented throughout much of the 19th century. The rugged topography was a major impediment to growth, but topography was only one of several characteristics that frustrated early developers. Another barrier to growth was the county's geographic position within the Bay Area. There was little reason for traffic to pass through the county because Marin was out of the way of major western migration paths and not on any route to the gold fields.

Mount Tamalpais and the adjoining stretch of the coastal mountain range dominate the geography of Marin. As a result, the county is divided, with the western portion separated from the eastern portion by the mountainous ridge. That geographical phenomenon has had a strong effect on Marin history. The county's towns, separated by peaks and valleys, developed in relative isolation and independence from one another.

Marin's pattern of land ownership was also an obstacle to growth and was frequently the subject of criticism in the late 1800s. The Spanish/Mexican land grant system proved to be more tenacious in Marin than in other parts of the state. Most of the land in the county was held by only a few families. As a result, Marin remained slow to develop relative to surrounding counties, particularly San Francisco, until after the 1906 earthquake.

A flood of weekend and vacation visitors to Marin was made possible by the development of ferry service, which began in the late 1850s, and later, rail transportation within the county, which grew during the late 1800s. The San Francisco business elite saw opportunity in that situation. Businessmen came to Marin to provide transportation systems, build resort hotels, develop residential communities, and provide the requisite public services, especially water. Private water development and land subdivision were the primary twins fostering growth in Marin. After the 1906 earthquake and the construction of the Golden Gate Bridge in 1937, county population growth created additional pressure for more water.

From the time of the Coast Miwok tribe, Mount Tamalpais provided the people of Marin with water. The earliest inhabitants simply relied on the many springs and creeks on the mountain. But as society became more complex, so did the technology and the systems for extracting the community's water supply. Unlike many areas in California, Marin has come to rely largely on local sources for its water supply, most of it generated from the Mount Tamalpais watershed. After the 1860s, as the population increased, it became necessary to build reservoirs to capture even more of the mountain's water. Today, the Marin Municipal Water District operates seven reservoirs within Marin.

During the 19th century, in order to promote growth, the California legislature passed laws encouraging the development of urban water companies. In Marin, government-sanctioned monopolies were created to provide water. Initially, local Marin residents, reflecting the frantic need for water, applauded the effort. But within a few years, local consumers began to see these same water companies as greedy monopolists. Over a 30-year period, the state legislature made several attempts to regulate the rates and services of the private water companies. Each of these attempts failed, sometimes because of political disagreement between the regulatory decision-makers, sometimes because of litigation, and sometimes because of the simple failure to act by dysfunctional local county boards or town governments.

Within the short period of a single lifetime, Marin underwent the process of recasting from a simple to a more complex society. Values began to change in some of the urban Marin communities. More and more citizens began to feel a failure of government. This was especially seen in the management of water resources.

The call for better water service and lower rates grew steadily in those parts of the community that were first to urbanize. The instability caused by having a dozen private water companies, the frustration over perceived high rates, and the failure to provide appropriate relief from poor service created a mounting call for public control of the water supply.

After the turn of the 20th century, with the emergence of the Progressive era, the old order began to give way in Marin townships. The Progressives had a penchant for conservation and a desire for the efficient management of natural resources, especially water resources. These attitudes began to make political change possible. Marin was reaching a political tipping-point. The perception grew that the private water companies, with their need for profit, were not suited to achieve the goal of an economical and reliable urban water supply.

The urbanized eastern part of Marin developed a strong body of Progressive leaders, including people like William Kent, George Harlan, and Michael M. O'Shaughnessy. Like Progressives throughout the country, the Marin Progressives were willing to experiment with governmental structure and to promote legislative changes in order to achieve their political goals. In 1910, when running for Congress, Kent acknowledged the east-west geographical division of Marin and the political and economic problems caused by the disparate urbanization in the county. He proposed to solve the problem by dividing Marin into two counties. The "two county proposal" failed, but the failure established the groundwork for other innovative governing structures. Two years later, Kent and other local Progressives persuaded the California legislature to implement legislation authorizing the use of "special districts" as regional, single-purpose, municipal water suppliers (the Municipal Water District Act of 1911). In April 1912, the urban voters of the eastern corridor of Marin approved the creation of the Marin Municipal Water District. In the following years, using authorized powers of eminent domain and a significant donation of land on Mount Tamalpais by William Kent, the district had taken the lands and waterworks of 24 privately owned companies and converted them to a single publicly owned system. The voter approval of 1912 came after a successful campaign that promised not only a water supply, but also a park. Since that promise, the Marin Municipal Water District has been charged with providing a reliable and pure water supply as well as public access to Mount Tamalpais.

One

Water and the Mountain in Early Marin

In the 1870s, Marin was a growing county. Early developers set up privately held water companies hoping to sell home sites. In 1871, William Tell Coleman set up such a company in San Rafael: the Marin County Water Company. The new company was established to facilitate the sale of real estate in Coleman's Magnolia Park (now the Dominican neighborhood in San Rafael), but also to supply other potential development in the San Rafael area and the existing state prison at San Quentin. Anticipating demand, Coleman's company built its first dam in 1872, creating Lake Lagunitas on the north side of Mount Tamalpais. The first of Marin's reservoirs delivered water through a combination of primitive pipes and redwood flumes from Lake Lagunitas to San Rafael. By 1880, through risky ventures and mismanagement, Coleman was in financial ruin. He was forced to sell his assets, and the Marin County Water Company was purchased by an English company, Falkner Bell & Co. It was managed by English-born James D. Walker until his death, when it was returned to local control. Prominent San Rafael resident and board member Arthur W. Foster gained control of the company. Bad service and the Marin County Water Company's need for profit left most citizens disgruntled.

In southern Marin, the Tamalpais Land and Water Company was created to provide water to promote the sale of real estate in Mill Valley. The company hired Michael O'Shaughnessy to survey and find water for the development of Mill Valley. In 1895, he completed Cascade Dam, creating the first significant water storage on the south side of Mount Tamalpais. O'Shaughnessy went on to play a major role in the development of the Marin Municipal Water District and the Hetch Hetchy water system for San Francisco.

By 1903, Lovell White, president of the Tamalpais Land and Water Company, separated land sales from the water business. That division created the North Coast Water Company, which devoted itself to increasing profits from water sales and, just as it had in San Rafael, this resulted in community dissatisfaction. Over the next decade, Marin citizens clamored for public control over their water supply.

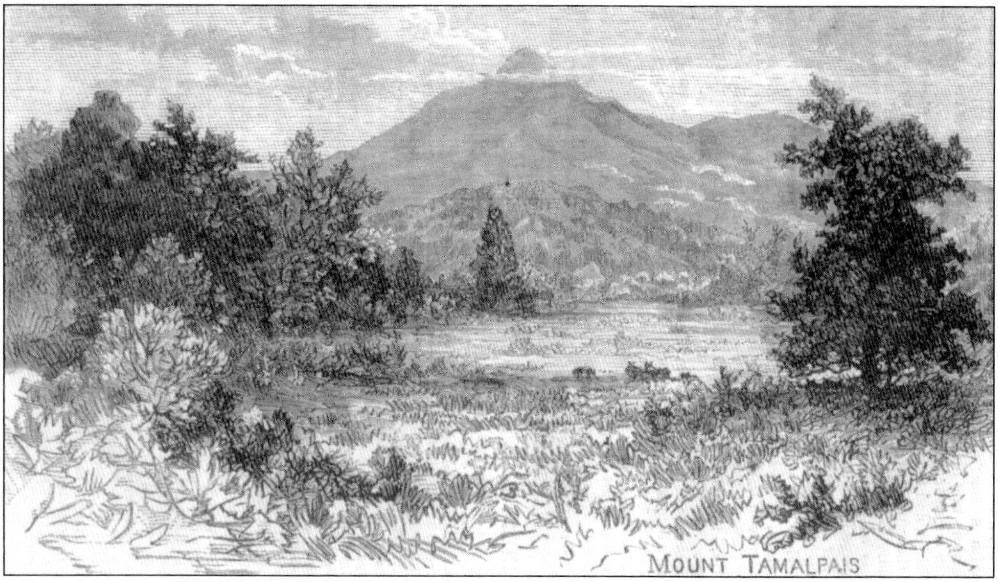

Tourists were attracted to the beauty of Mount Tamalpais as early as the 1870s. In May 1875, *Harper's Weekly*, a national publication, presented an article describing tourist attractions in the San Francisco area. *Harper's* commissioned artists Jules Tavernier and Paul Frenzeny to travel the West and create art for illustration in the weekly. Their woodcut of Mount Tamalpais (above) was featured in the 1875 article.

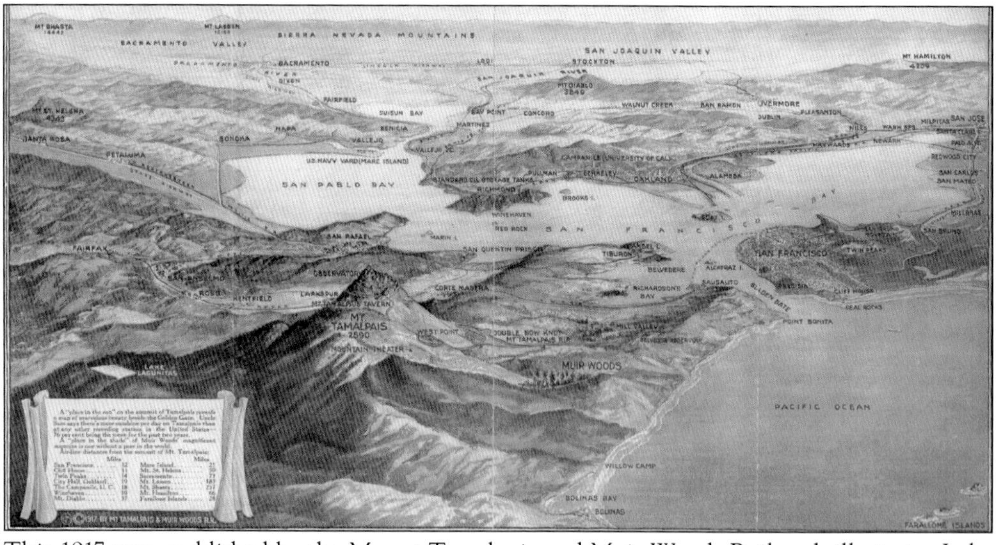

This 1917 map, published by the Mount Tamalpais and Muir Woods Railroad, illustrates Lake Lagunitas, the Mountain Theater, Muir Woods, and other tourist features on Mount Tamalpais. These are set against the background of Northern California, from Mount Shasta in the north to Mount Hamilton in the south. Tourism on Mount Tamalpais increased heavily during the period partly due to this promotional publicity. (Courtesy of the Anne T. Kent California Room, Marin County Free Library.)

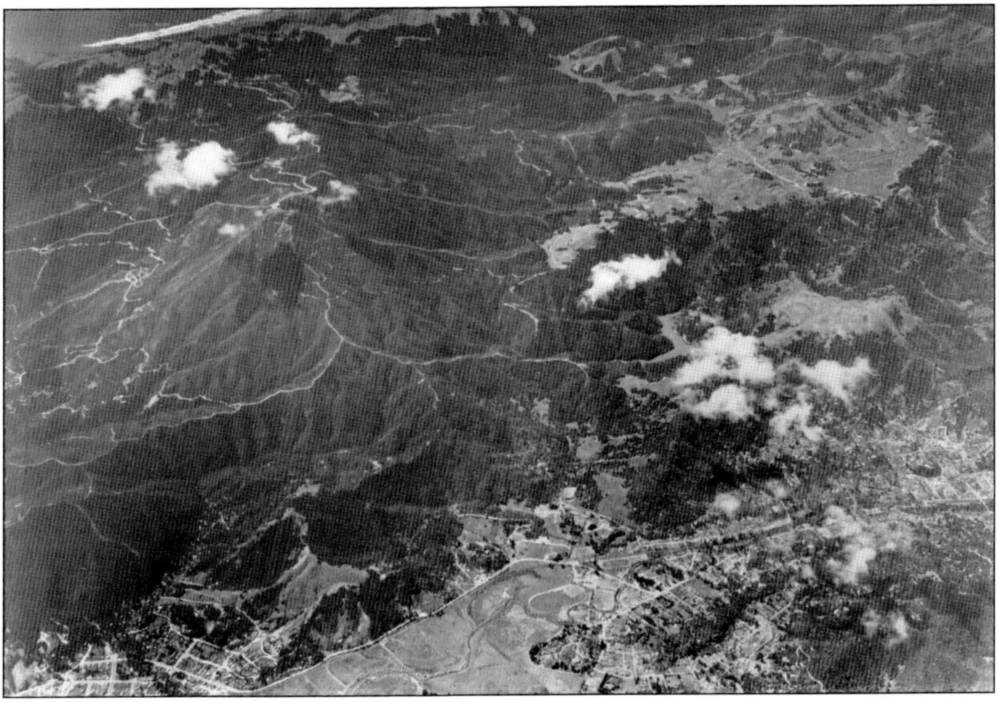

These two aerial photographs of Mount Tamalpais were taken from similar perspectives, looking toward the mountain in a westerly direction. The above photograph was taken in 1930 and the below photograph was taken in 1954. During that period of time, the population of Marin County more than doubled, yet the mountain and its surrounding watershed remained in its undeveloped condition, as a result of the lands coming under public control with the creation of the Marin Municipal Water District and Mount Tamalpais State Park.

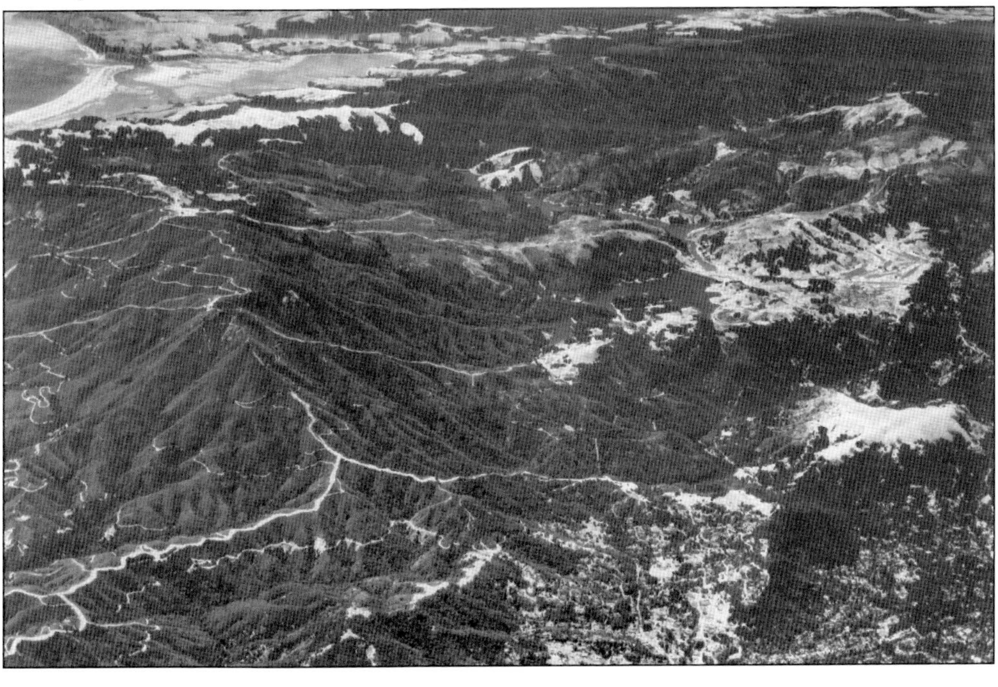

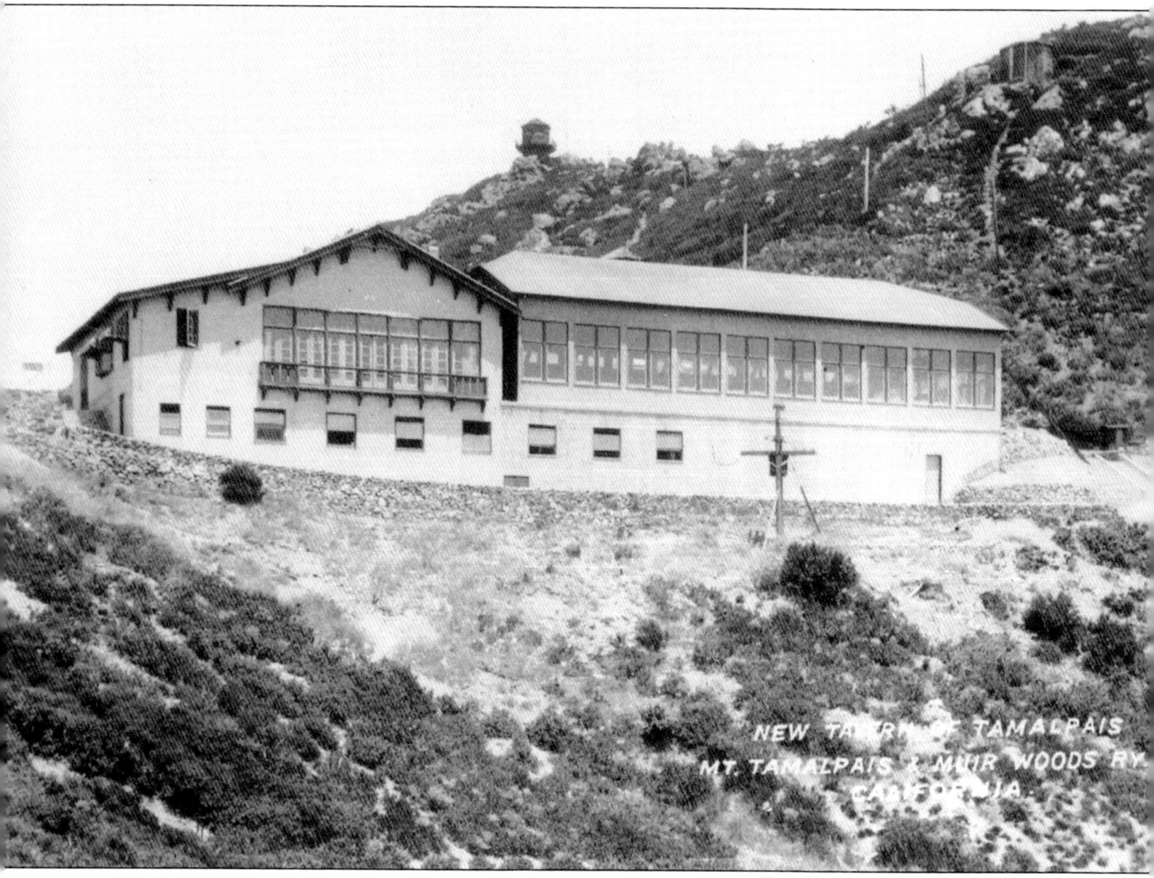

The Mill Valley & Tamalpais Scenic Railway was designed to bring tourists to Mount Tamalpais. As it was being completed in 1896, the Tavern of Tamalpais was built as an added attraction at the end of the track, 200 feet below the summit. It was advertised that on a clear day, the Sierra Nevada Range could be seen from the tavern's veranda. This was the end of the 8.2-mile railway, which began in Mill Valley and ended at the tavern (seen here in 1913). MMWD acquired the tavern in 1930, when the railroad went out of business. Prior to that, the tavern was owned by the Tamalpais Land and Water Company and its successor, the North Coast Water Company. The land was later conveyed to Mount Tamalpais State Park in 1948. The tavern was razed in 1950.

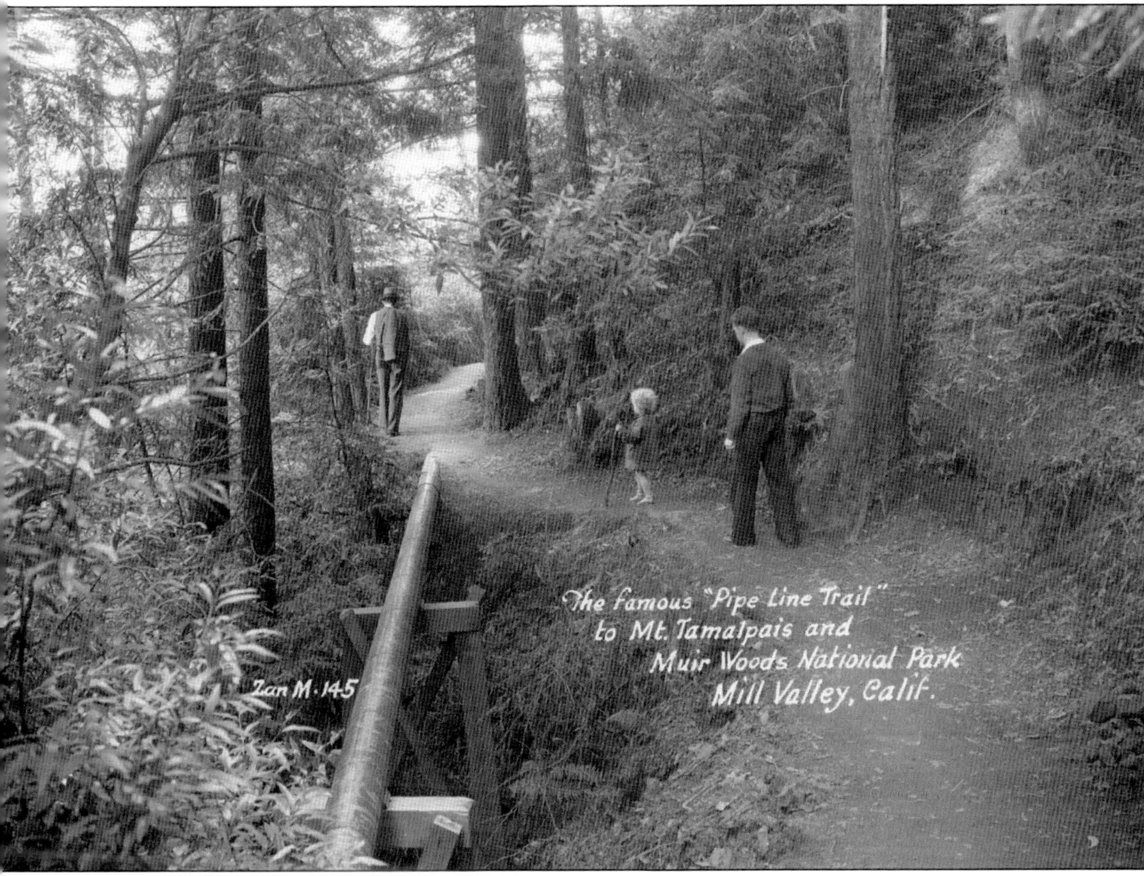

In the early part of the 20th century, as the populations of Marin's towns were growing, the struggle for an adequate water supply grew more intense. The Tamalpais Land and Water Company and its successor, the North Coast Water Company, had developed a water supply for Mill Valley, a newly created community developed by the Tamalpais Land and Water Company. Fierce competition grew between the towns of Sausalito and Belvedere for a limited water supply. Both towns made bids for water from an expanded system to be provided by the North Coast Water Company. In 1904, Belvedere outmaneuvered Sausalito when the Belvedere Land Company signed an agreement to receive water from the North Coast Water Company. As a result of the agreement, the water company constructed a 12,000-foot line of eight-inch steel pipe along a trail on the south side of the mountain—today's "Pipeline Trail." The source of the water was the north fork of Laguna Creek. (Courtesy of the Anne T. Kent California Room, Marin County Free Library.)

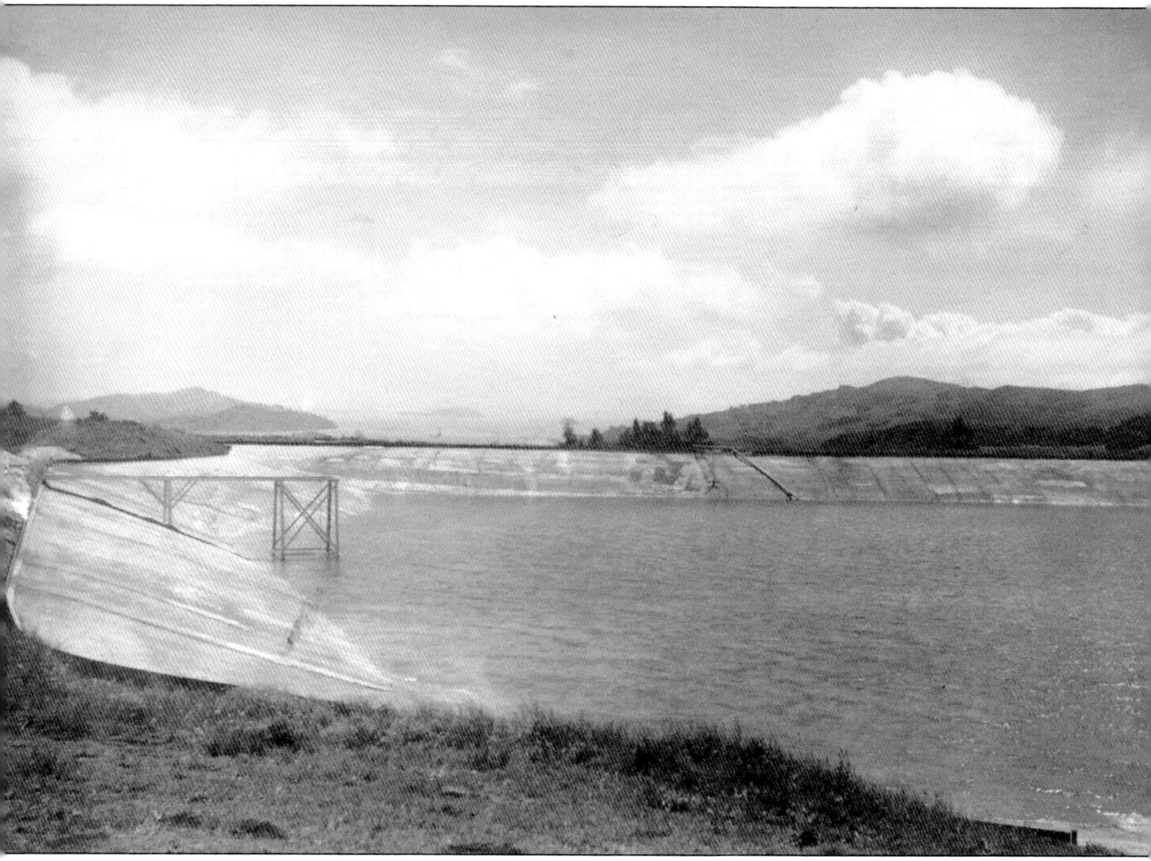

The Belvedere Dam and its reservoir were built by the North Coast Water Company in 1905 on a nearly 10-acre parcel of land purchased for $693.70, above what is now Sequoia Valley Drive and Edgewood Avenue in Mill Valley. The dam consisted of an indentation in the hills in which water was stored by means of two riprap earthen dams, resulting in a capacity of 8.3 million gallons. The Belvedere Land Company purchased stock from the North Coast Water Company in order to obtain the water. Early on, the dam developed leaks, caused by gophers digging into the outside face of the dam. Leaks amounted to 70,200 gallons per day in 1911, of which only 43,200 could be collected and returned to the system. When MMWD installed facilities for Belvedere and Tiburon in 1921, the name Belvedere Dam gradually changed to Mill Valley Reservoir. It was taken out of service in 1967.

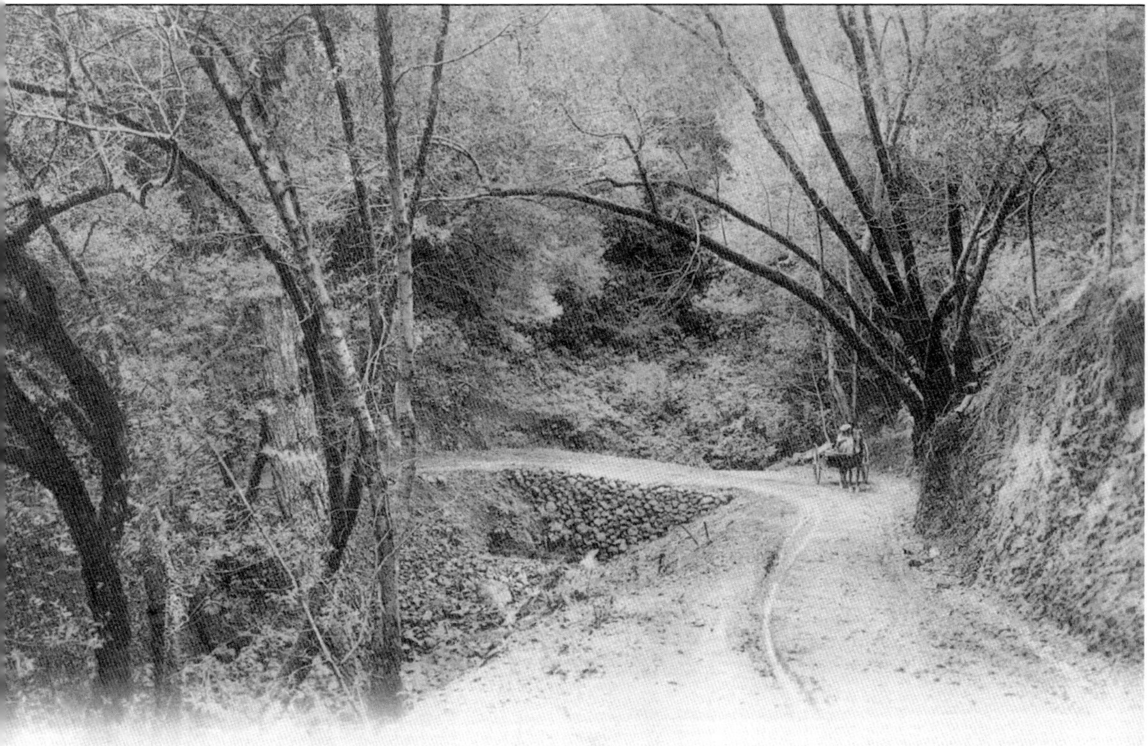

17 The Wagon Road, Mt. Tamalpais, California.

Since the pioneer days in Marin, there has been interest in building a carriage road to the top of Mount Tamalpais. Early trails went past Blithedale Canyon, William Kent's property, and Lagunitas Lake. When the Ross-to-Bolinas road was built in 1878, many Marin residents advocated combining that project with construction of a road to the top of the mountain. In 1881, John C. Eldridge, a successful San Francisco merchant and owner of the San Rafael Gas Works, undertook to build a road to the summit of the west peak. By August 1884, Eldridge had raised half the money and began construction, using more than 100 Chinese laborers. In December 1884, the road—just over seven miles—was completed, and on Saturday, December 14, three two-horse carriages brought dignitaries from San Rafael to the summit for opening ceremonies. The road remained popular for only a few years, losing its glitter when the Mountain Railroad was built later in the decade. The lower road was abandoned in 1905, when the Marin Water & Power Company constructed Phoenix Dam and Reservoir.

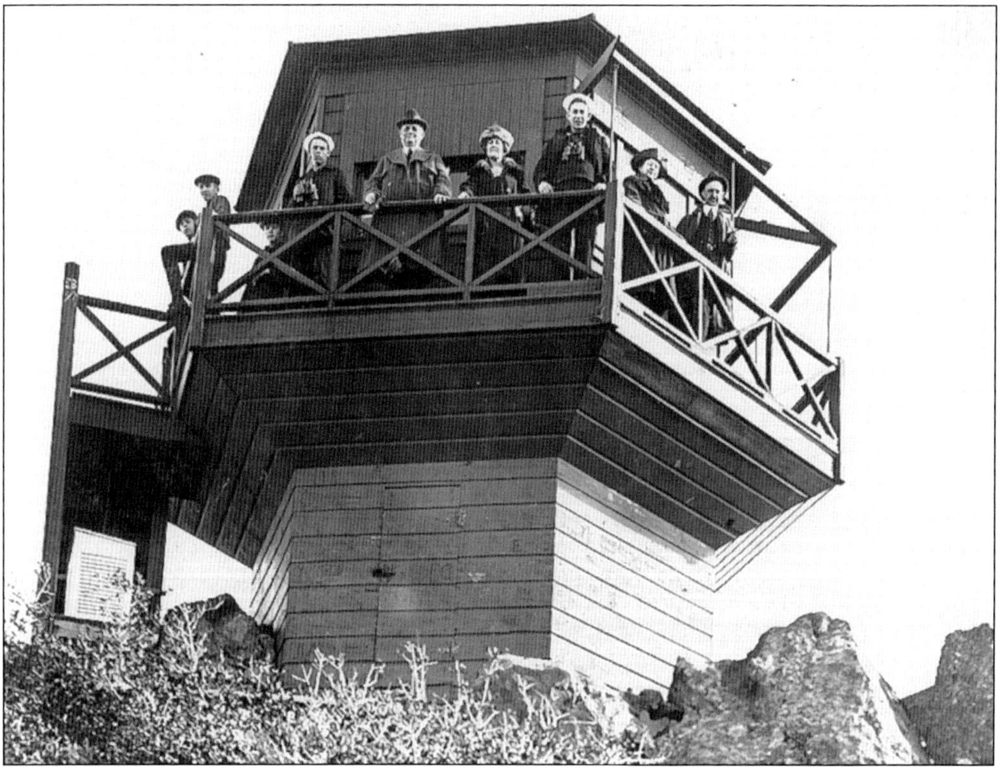

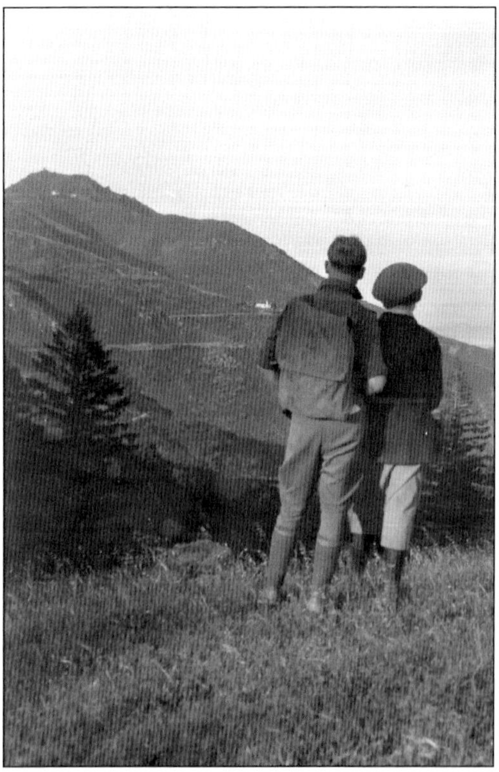

The lookout station on the east peak of Mount Tamalpais was built by the *San Francisco Examiner* in 1901 to spot ships approaching San Francisco. It became a fire lookout in 1921 and later a popular destination for hikers and tourists, seen in this photograph from about 1922. In 1935, the Civilian Conservation Corps rebuilt the aging structure, which was subsequently named Gardner Lookout, after fire chief Edwin B. Gardner. (Courtesy of the Lucretia Little History Room, Mill Valley Public Library.)

By carriage, wagon, and horseback, but mostly on foot, visitors passed daily over privately owned dairy lands and open grasslands. Many hiking and outdoor clubs were formed—the Tamalpais Club (1880), the Sightseers Club (1887), the Cross-Country Boys Club (1890), John Muir's Sierra Club (1892), and the Tamalpais Conservation Club (1912)—and all promoted activities on Mount Tamalpais. (Photograph by Lothers and Young Studios, courtesy of the Anne T. Kent California Room, Marin County Free Library.)

Hikers have long been attracted to the summit of Mount Tamalpais. Before the turn of the 20th century, tourists were frequently found on the mountaintop dressed in fine fashion, which the rugged trails on the mountain did not accommodate. But after 1896, with the help of the Mount Tamalpais Scenic Railway (later the Mount Tamalpais & Muir Woods Railway) running from Mill Valley to the top of the mountain, access to the mountain peak became much easier. Once near the east peak, well-dressed sightseers were aided by the Plank Walk Trail, a boardwalk from near the Tavern of Tamalpais to the east peak that facilitated more urban adventurers. (Photographs courtesy of the Lucretia Little History Room, Mill Valley Public Library.)

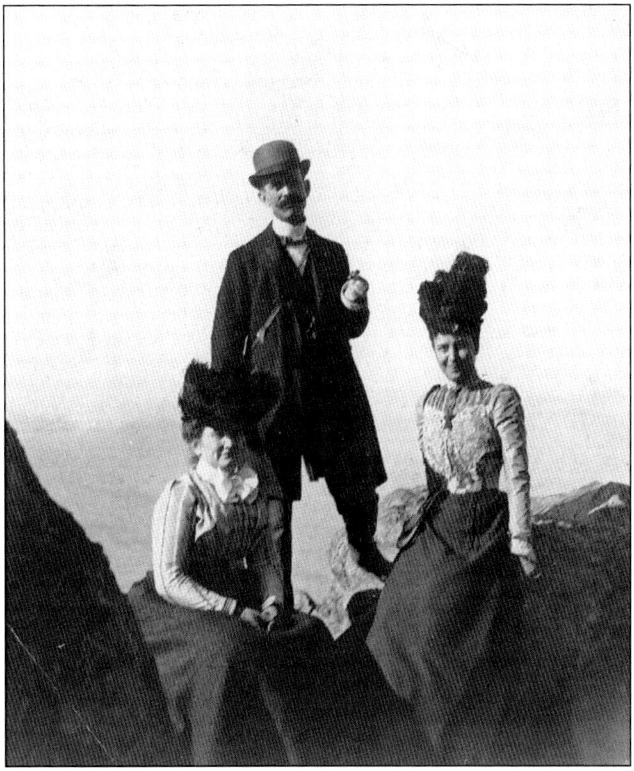

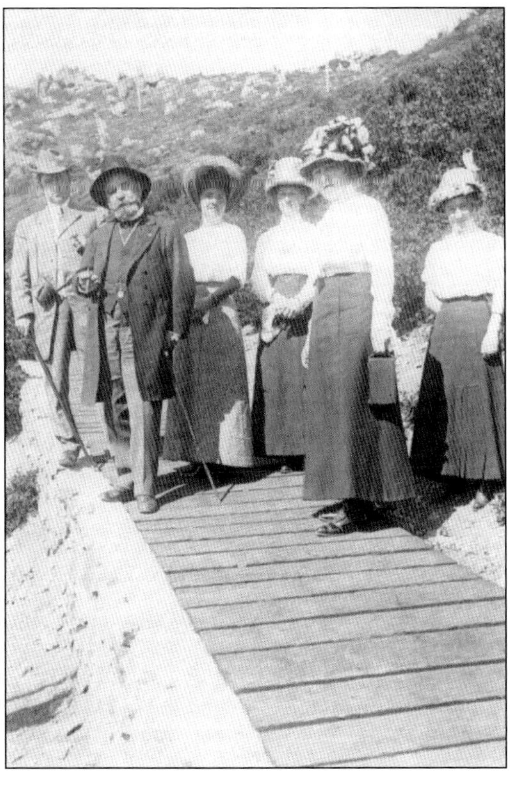

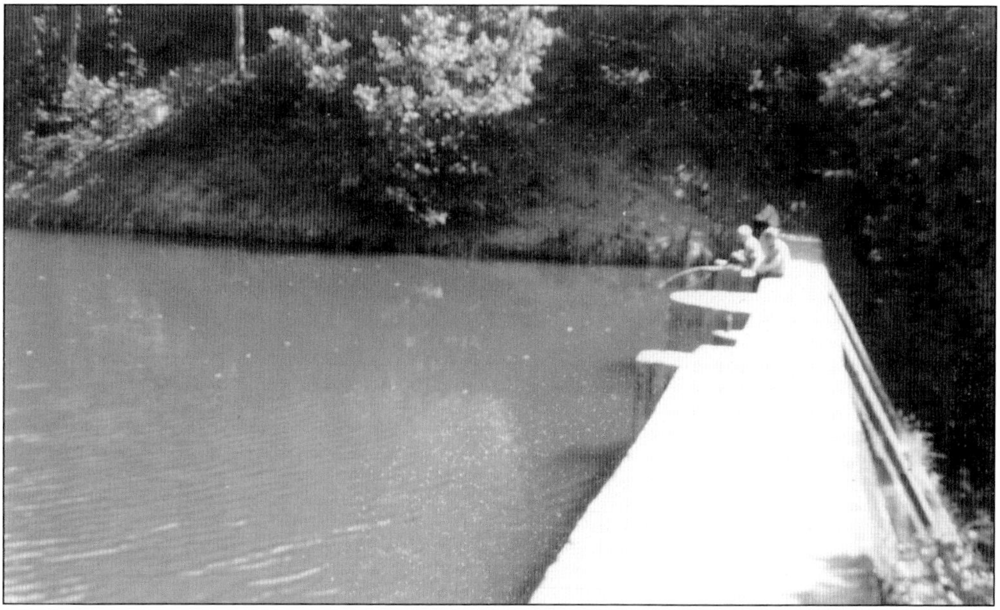

The Tamalpais Land and Water Company hired engineer Michael O'Shaughnessy to find water for the development of Mill Valley. In 1895, the Mill Creek Dam (later called the Cascade Dam) was completed, creating the first water storage on the south side of Mount Tamalpais. The dam was taken out of service in 1956, and the City of Mill Valley took it over in 1959 for fire protection services.

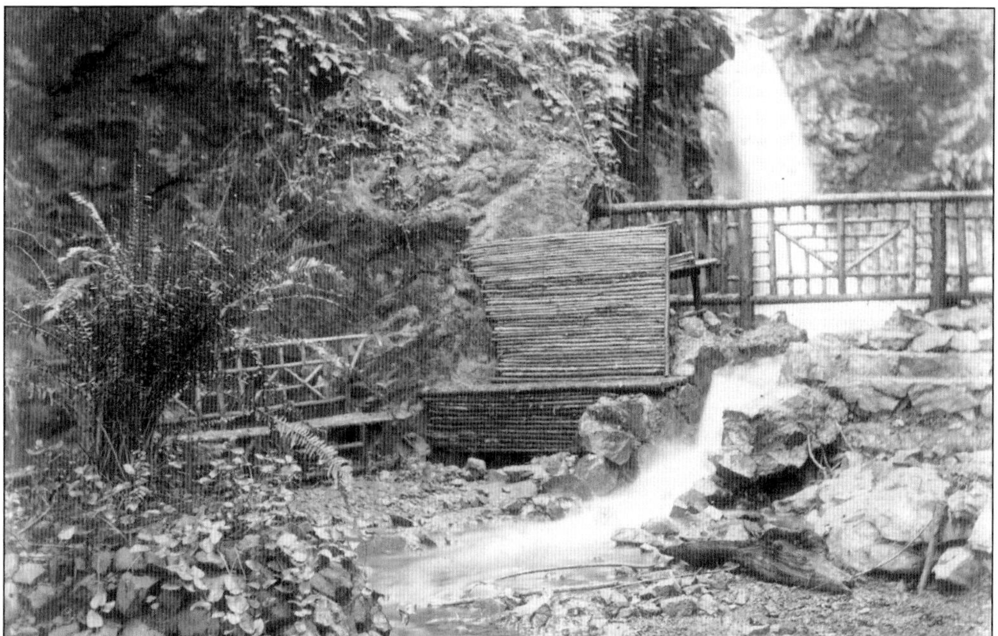

The Outdoor Art Club, an outgrowth of the early-20th-century Women's Club movement, under the leadership of Laura White and other club leaders, contributed many decorated benches, rails, and bridges throughout the southern watershed areas. This example, which dates to the early 1900s, was constructed in the Cascade Falls vicinity. (Courtesy of the Lucretia Little History Room, Mill Valley Public Library.)

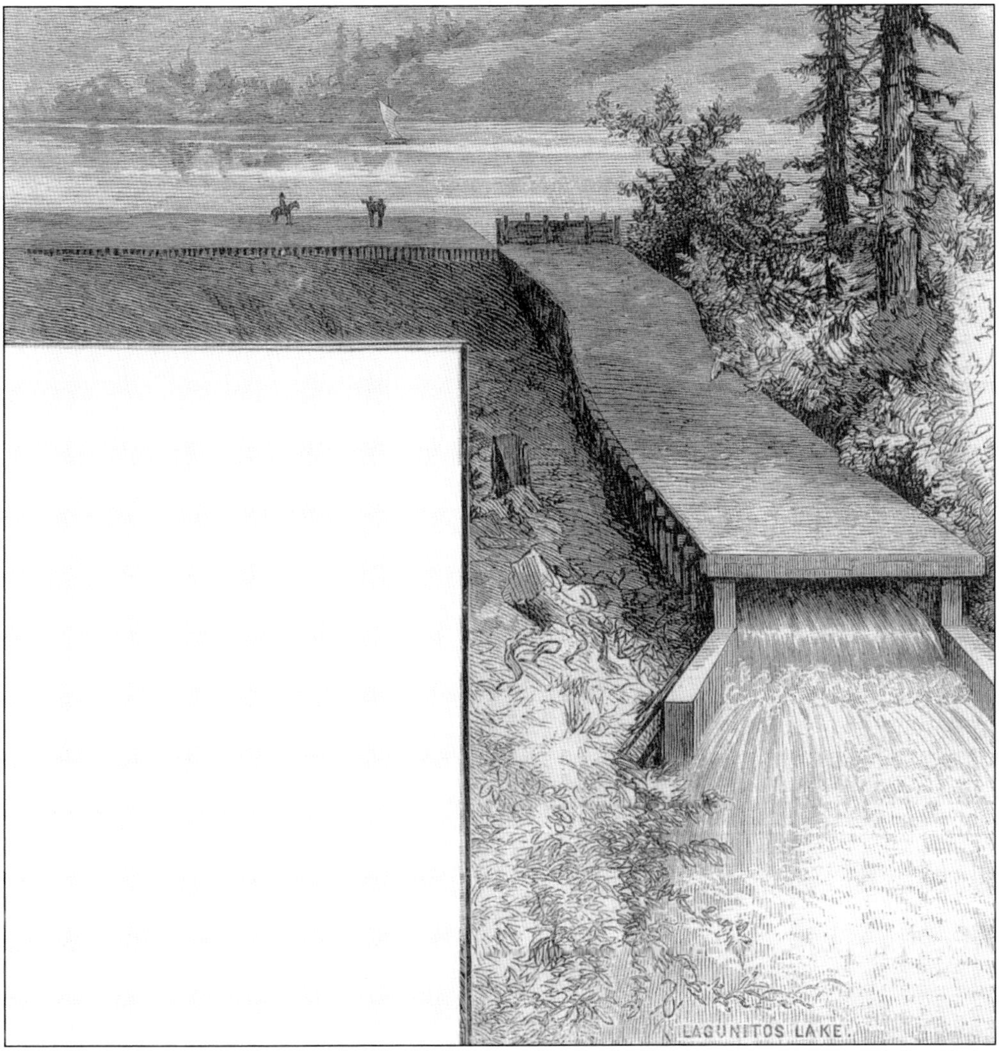

William Tell Coleman hired Herman Schussler, the engineer for San Francisco's Spring Valley Water Company, to construct a dam, reservoir, and transmission system to bring water from Lagunitas Creek to San Rafael. Designed as a gravity system, the reservoir fed water into an open wooden flume (depicted in this early *Harper's Weekly* illustration from 1875), which had an elevation drop of 100 feet per mile. This rate compared favorably with the best section of the Spring Valley system, which had a fall of 42 feet per mile, and the slower 13 inches per mile of the Croton waterworks, which supplied New York City. On the day the new transmission system was put into service, workers stationed at various levels along the transmission line emptied 50 sacks of bran and sawdust into the system to fill any small holes or leaks. Schussler's design included the dam and reservoir at the 780-foot level on the side of Mount Tamalpais. When completed, the reservoir held 150,000 gallons, a number that could be doubled by raising the dam only a few feet.

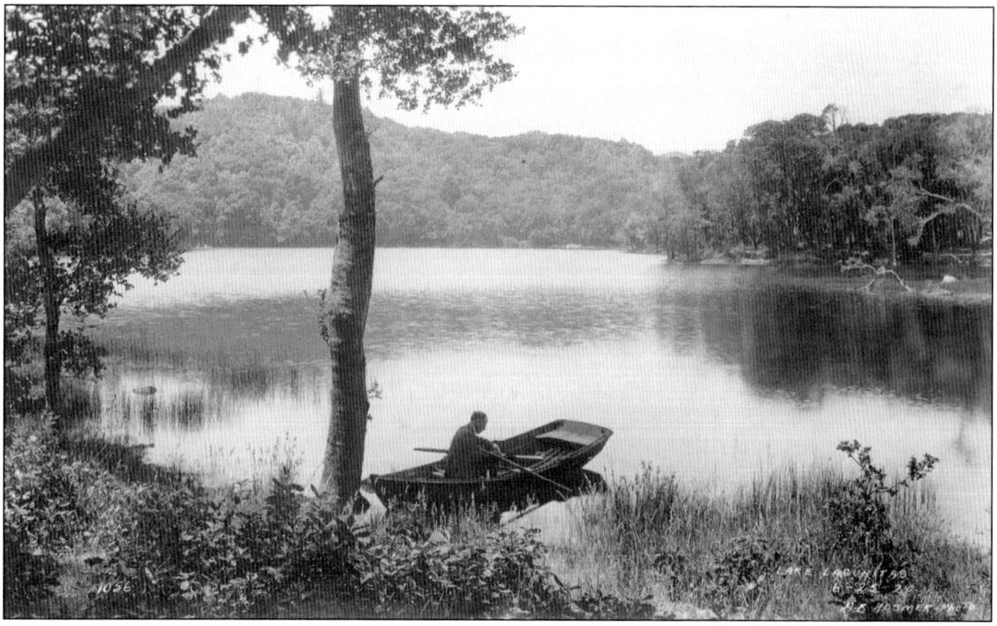

When Coleman fulfilled his promise to build Lagunitas Dam in 1873, he was already known as the chairman of the controversial San Francisco Committee of Vigilance. The dam was constructed by Chinese laborers who were paid $20 per month. An earth-filled dam, it was 480 feet long and 200 feet wide at its base. In California, at that time, only the Pilarcitos and San Andreas Dams, in San Mateo County, were larger. (Photograph by H.B. Hosmer.)

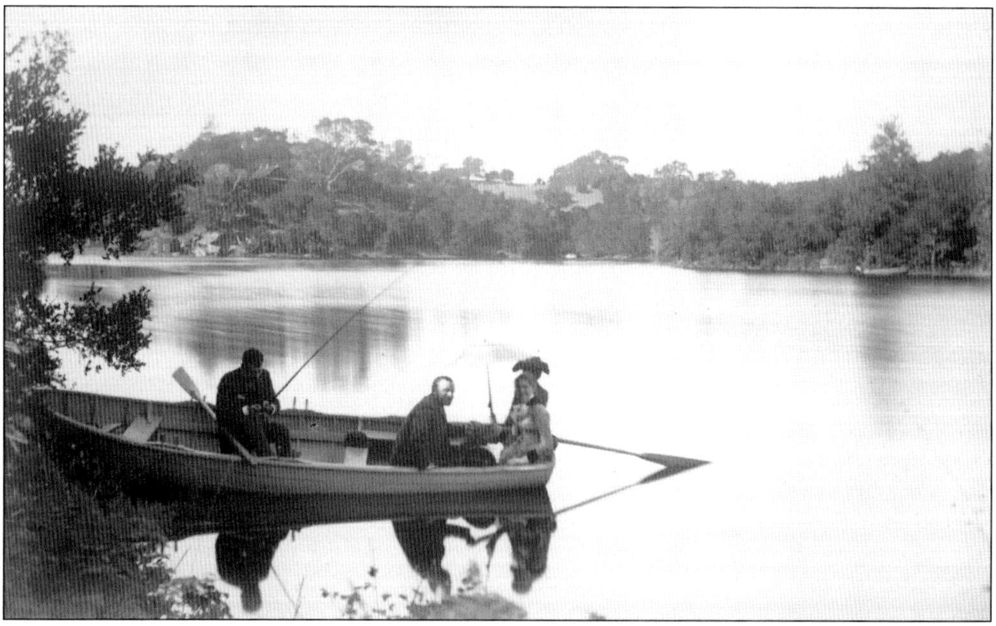

Before the dam was constructed, the area consisted of several ponds, thus the name Lake Lagunitas—Spanish for "little lakes." In 1873, it was Marin's only reservoir. Over the years, the district added significant amounts of additional reservoir storage and the importance of Lake Lagunitas as a source of water diminished. Excepting the drought years of 1976–1977, it has not put water into the system since 1963. (Courtesy of the Bancroft Library, University of California, Berkeley.)

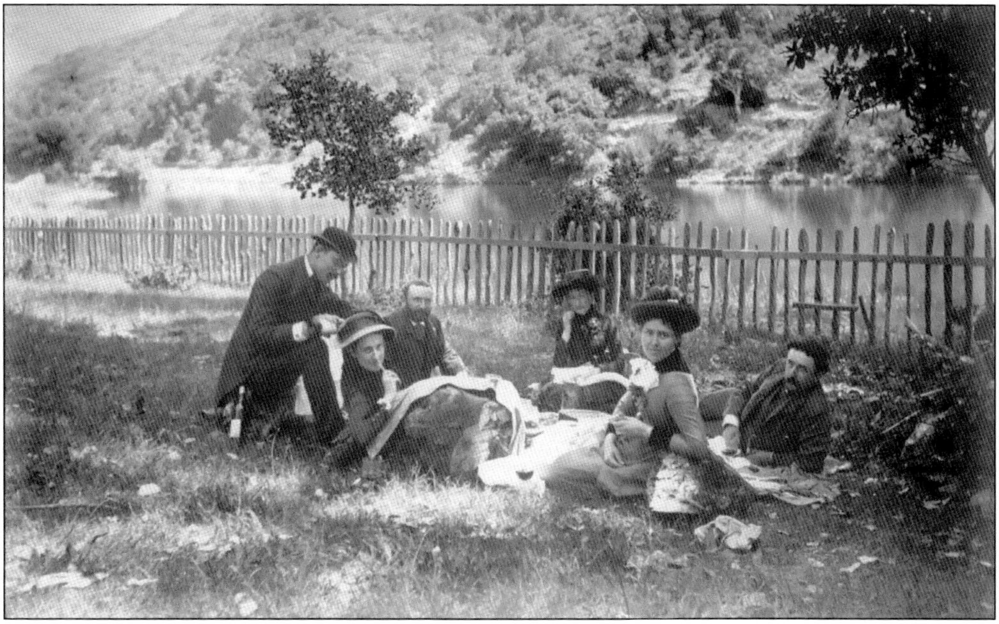

Picnickers and hikers have always been attracted to Lake Lagunitas. During the Depression, the New Deal Civilian Conservation Corps (CCC) built or improved many of the picnic areas, trails, fire roads, and other facilities on Mount Tamalpais that are still frequently used today. (Courtesy of the Bancroft Library, University of California, Berkeley.)

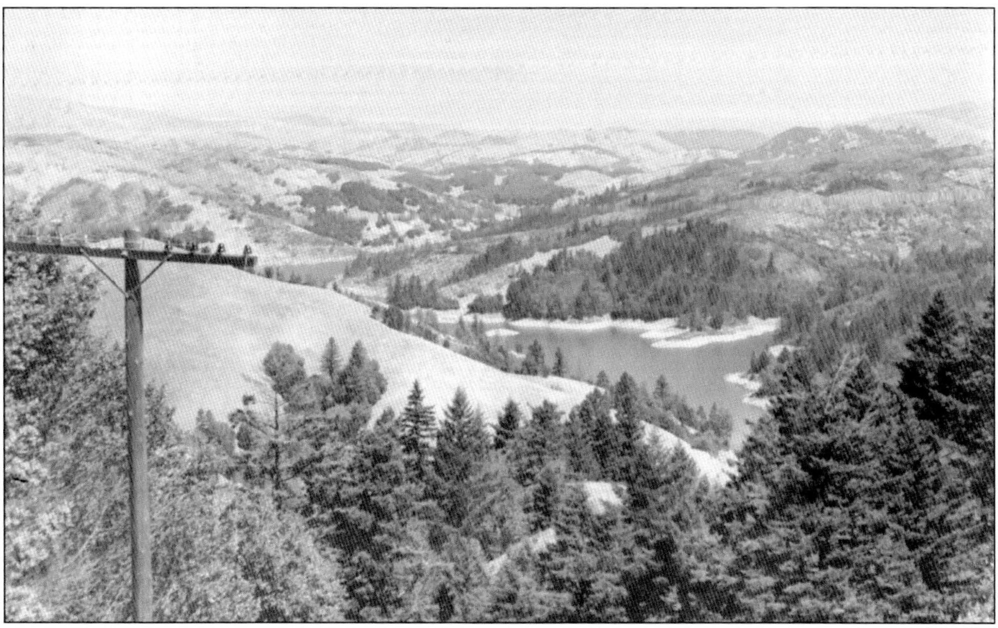

Marin's first telephone line was completed in 1877 and ran from the Marin Water Company office on C Street in San Rafael to the dam keeper's house at Lake Lagunitas. The *Marin Journal* reported, "The Water Company's telephone is a perfect success, and . . . many citizens drop in at the office to hear Mr. Fox converse from his residence at the dam." Sharp-eyed visitors to the watershed can spot remains of the line from Lake Lagunitas to Phoenix Lake. (Courtesy of the Anne T. Kent California Room, Marin County Free Library.)

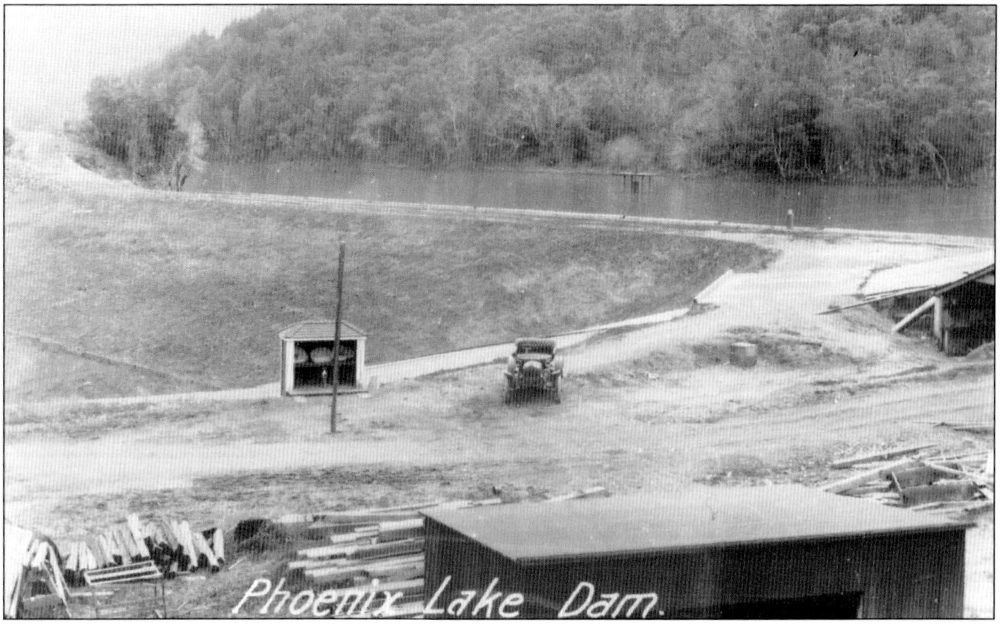

By 1904, Marin was growing, which caused pressure on the water supply in all of the towns. The Marin Water & Power Company hoped to build a new dam, identified as Tamalpais Dam, in the vicinity of today's Alpine Dam. Faced with controversy over the location as well as litigation with the county, the Tamalpais Dam was never built. Arthur W. Foster, the president of the company, described the situation as a water emergency. The emergency was caused by a water shortage, but it was aggravated by Foster's race with the North Coast Water Company to reach the growing water market developing in the southern part of the county (Sausalito and Belvedere). Weighing these factors, Foster decided to abandon plans for the Tamalpais Dam and instead undertook a project at a less controversial location, building the dam in Phoenix Gulch, upstream of the community of Ross.

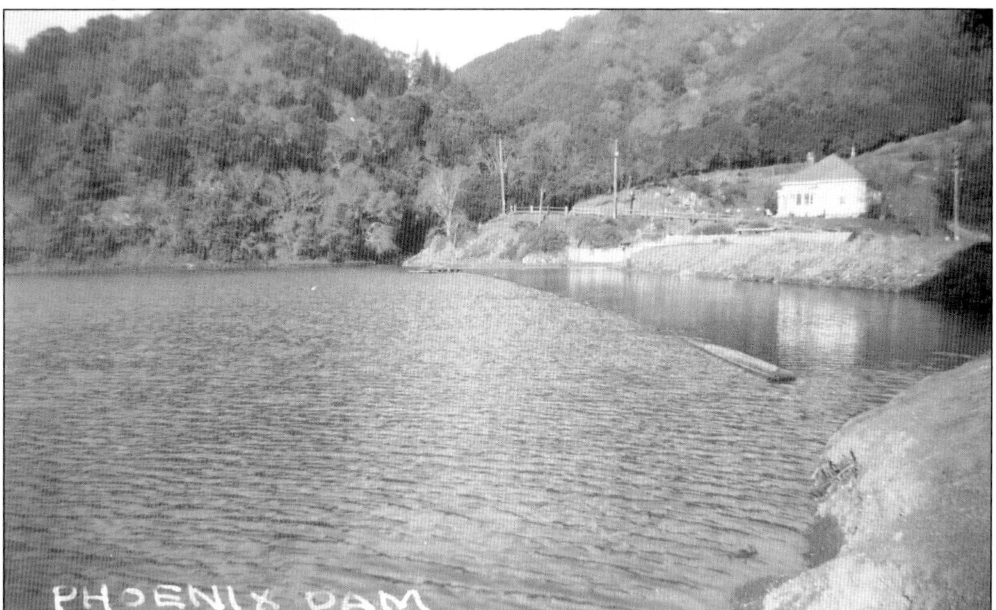

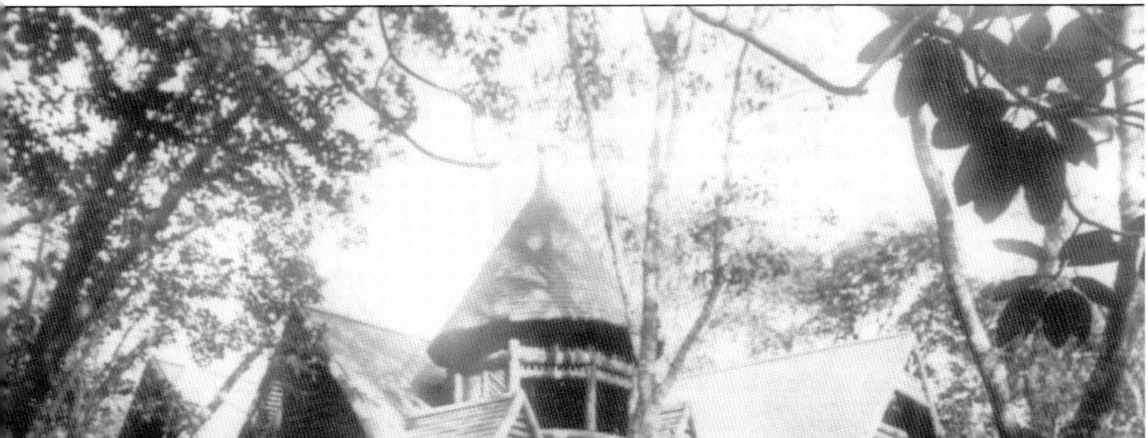

One of the architectural gems of MMWD lands is the old log cabin on the banks of Phoenix Lake. The log cabin sat on the Porteous property when it was built about 1894. James Stoddard Porteous purchased over 1,000 acres of Rancho Punta de Quentin in 1883. He and his wife, Janet Porteous, used the land for ranching—it was known as Hypolite Ranch. After James died in 1887, Janet lived on the property and built the log cabin as a ranch caretaker's house. The cabin's redwood logs were cut on the Deer Park section of the ranch, because Janet did not want any trees removed near the main house, which was uphill of the cabin, on the site where the water district's ranger house sits today. The curly redwood trim around the windows and doors came from a fallen tree on the southwestern side of Shaver Grade. The property was acquired by MMWD in 1916. This photograph dates to the late 1890s.

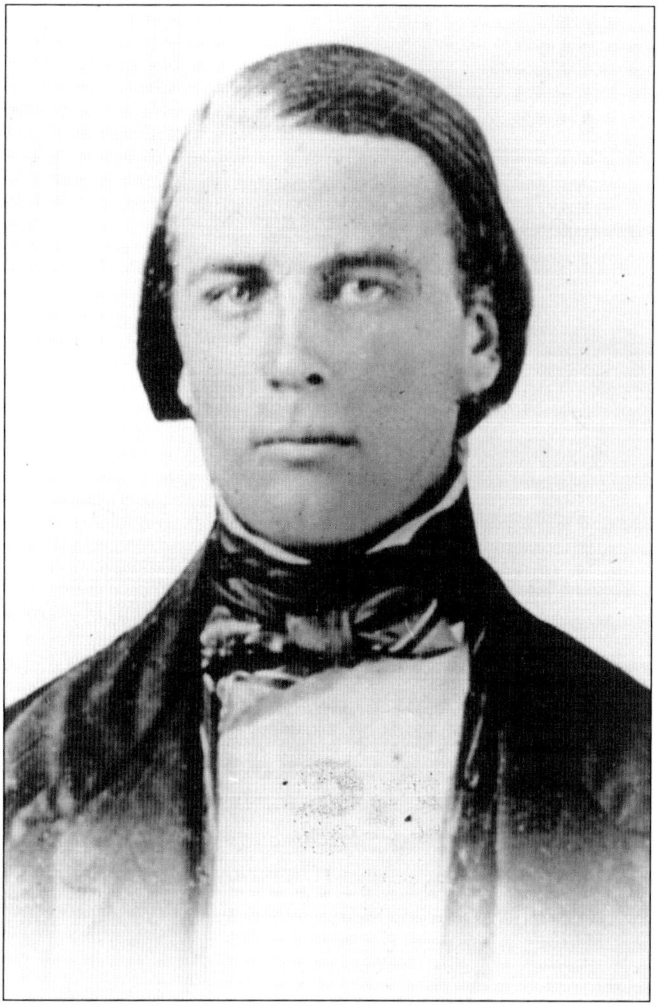

William Tell Coleman was a wealthy shipping merchant in San Francisco and the chairman of the controversial San Francisco Committee of Vigilance. He built his summer and weekend home at 1130 Mission Avenue in San Rafael. He was one of the first of a new breed of water entrepreneurs in Marin. Coleman realized that San Rafael was appealing to wealthy San Francisco businessmen as a location for summer or even permanent homes. He purchased over 1,000 acres in San Rafael called Irwin Ranch and renamed it Magnolia Park (the Dominican area today). He hired William Hammond Hall, the designer and first superintendent of San Francisco's Golden Gate Park, with the plan of designing a high-end residential community. Coleman knew its success depended upon providing a domestic water supply. (Photographs courtesy of the Marin History Museum.)

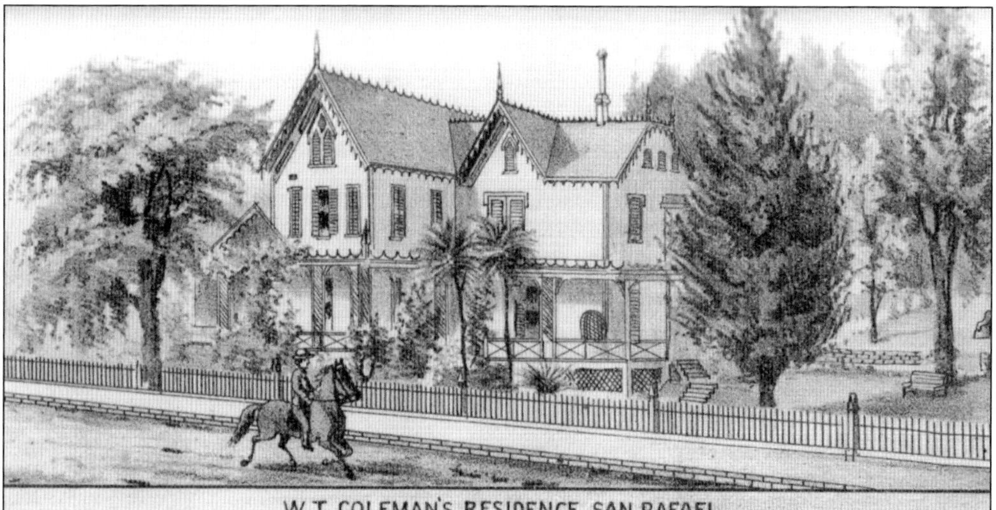

W.T. COLEMAN'S RESIDENCE, SAN RAFAEL.

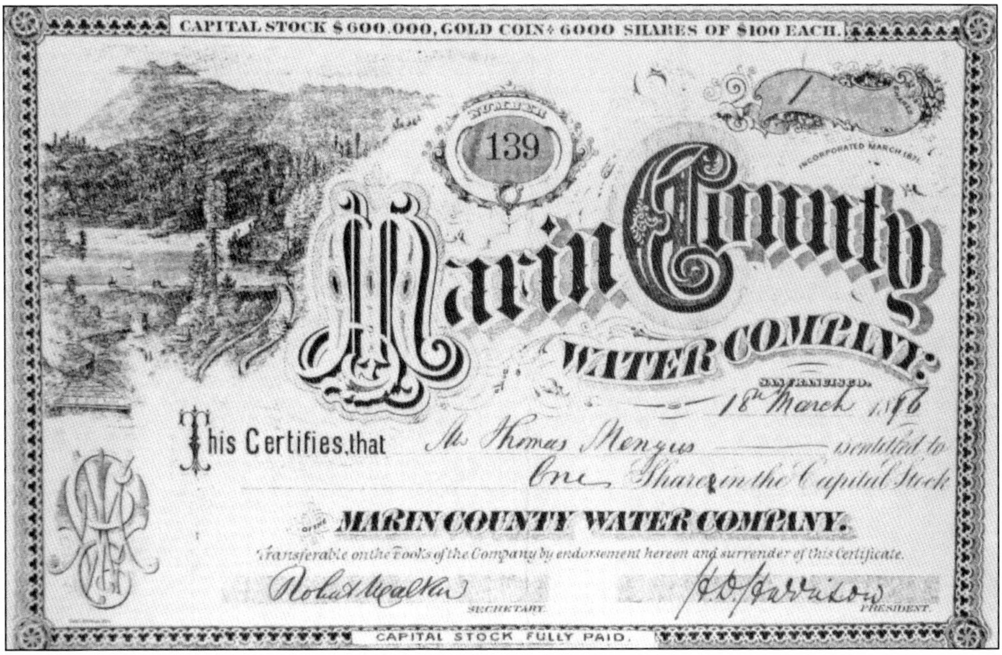
CAPITAL STOCK $600,000, GOLD COIN: 6000 SHARES OF $100 EACH.

139

Marin County Water Company.

INCORPORATED MARCH 1871.

SAN FRANCISCO, 18th March 1876

This Certifies, that Mr Thomas Menzies entitled to One Share in the Capital Stock of the MARIN COUNTY WATER COMPANY.

Transferable on the Books of the Company by endorsement hereon and surrender of this Certificate.

SECRETARY. PRESIDENT.

CAPITAL STOCK FULLY PAID.

Coleman could see that there were two obstacles to making profit from San Rafael real estate—access and water. But transportation was rapidly improving by the 1870s. The ferry charge was reduced from $2 to $1, with a monthly commute book priced at $15, and the San Rafael Stage Company reduced its fares from $1 to 50¢. This resulted in San Francisco businessmen spending more time in Marin County. With improved transportation, Coleman had only to address the problem of water. In 1871, he announced his intention to bring water to San Rafael from Lagunitas Creek. He incorporated the Marin County Water Company in 1872 and raised $300,000 for capital stock. Utilizing laws designed to encourage domestic water suppliers, he filed condemnation proceedings against several owners of large parcels on Mount Tamalpais, seeking to acquire rights to the creek and other streams on the watershed.

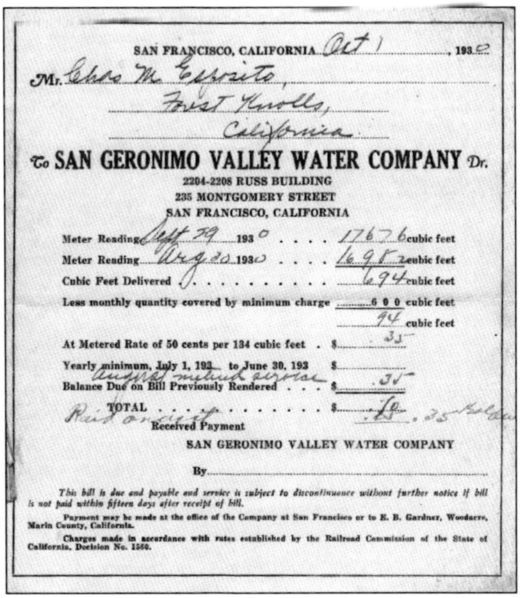
SAN FRANCISCO, CALIFORNIA Oct 1, 1930

Mr. Chas M. Esposito, Forest Knolls, California

To SAN GERONIMO VALLEY WATER COMPANY Dr.

2204-2208 RUSS BUILDING
235 MONTGOMERY STREET
SAN FRANCISCO, CALIFORNIA

Meter Reading Sept 29 1930 . . . . 17676 cubic feet
Meter Reading Aug 30 1930 . . . . 16982 cubic feet
Cubic Feet Delivered . . . . 694 cubic feet
Less monthly quantity covered by minimum charge 600 cubic feet
94 cubic feet
At Metered Rate of 50 cents per 134 cubic feet . $ .35
Yearly minimum, July 1, 193 to June 30, 193 $
Balance Due on Bill Previously Rendered . . . $ .35
TOTAL . . . . $ .70

Received Payment
SAN GERONIMO VALLEY WATER COMPANY

By

This bill is due and payable and service is subject to discontinuance without further notice if bill is not paid within fifteen days after receipt of bill.

Payment may be made at the office of the Company at San Francisco or to E. B. Gardner, Woodacre, Marin County, California.

Charges made in accordance with rates established by the Railroad Commission of the State of California. Decision No. 1560.

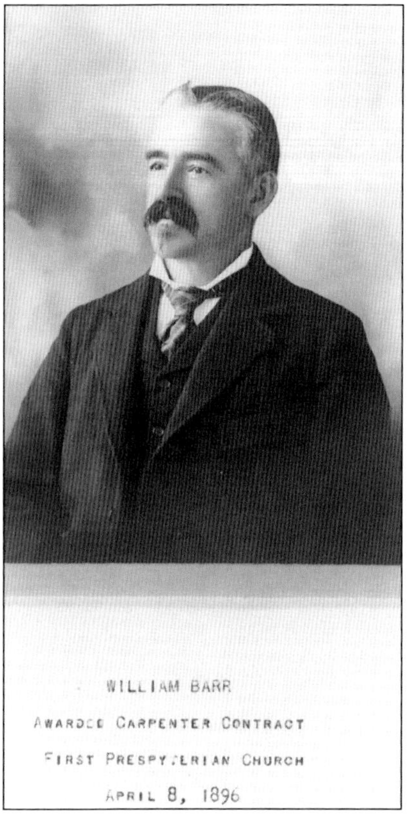

In the late 19th and early 20th centuries, William Barr was a very active member of the Marin community. Under the company presidency of Arthur W. Foster, Barr served as the superintendent of the Marin Water & Power Company—the equivalent of general manager. With Foster, Barr was instrumental in the creation of the seminary in San Anselmo, and he also served as a Marin County supervisor.

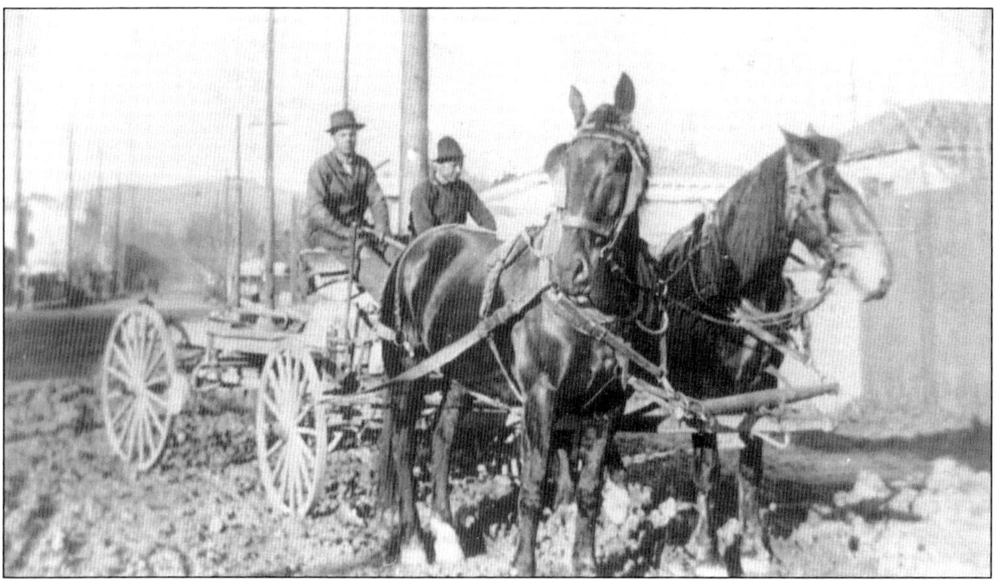

The Marin Water & Power Company needed an open yard for construction assembly and the storage of pipes, vehicles, and equipment. Shown here are two Marin Water & Power Company employees at the company's pipe yard at the foot of Petaluma Avenue in San Rafael (now Lincoln and Second Streets). That yard became the first pipe yard for MMWD after it acquired the company's assets in 1916.

In 1877, James D. Walker became president of the Marin County Water Company. Coleman, faced with financial difficulties, was forced to sell the company to British firm Falkner, Bell & Co. Coleman resigned, and the British-born Walker (pictured) became president. He lived in San Rafael during his time with the company, and, in 1897, died of a heart attack during a contentious town trustees meeting. (Courtesy of the Marin History Museum.)

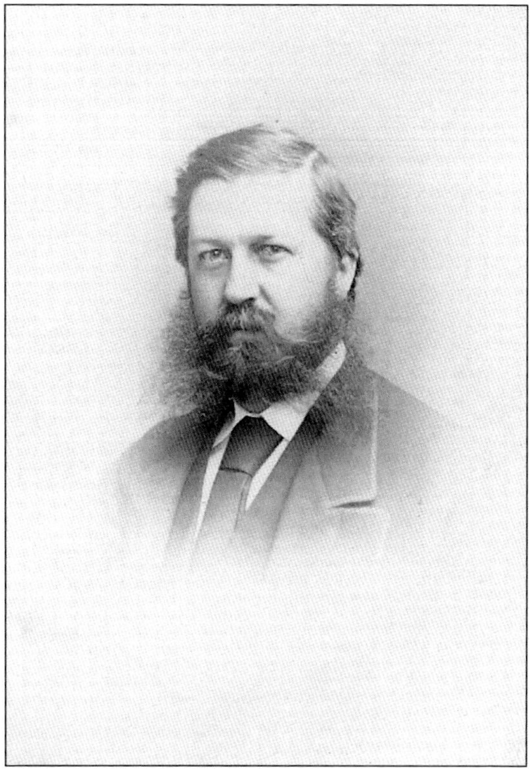

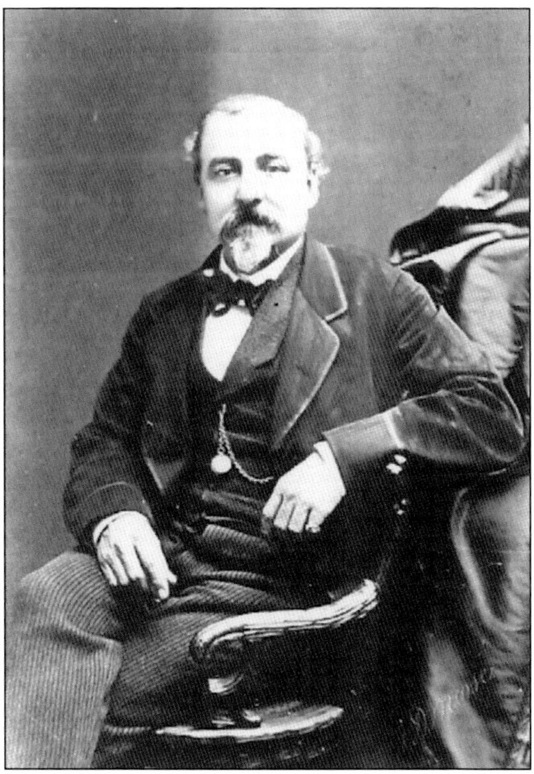

James Eastland was president of the San Francisco Savings Union when the bank foreclosed on Samuel Throckmorton's Rancho Sausalito. Eastland, with other members of the bank's board of directors, created the Tamalpais Land and Water Company and bought the property for development. Using Eastland's wealth, the group created the town of Mill Valley, which was, for a short period, called Eastland. (Courtesy of the Marin History Museum.)

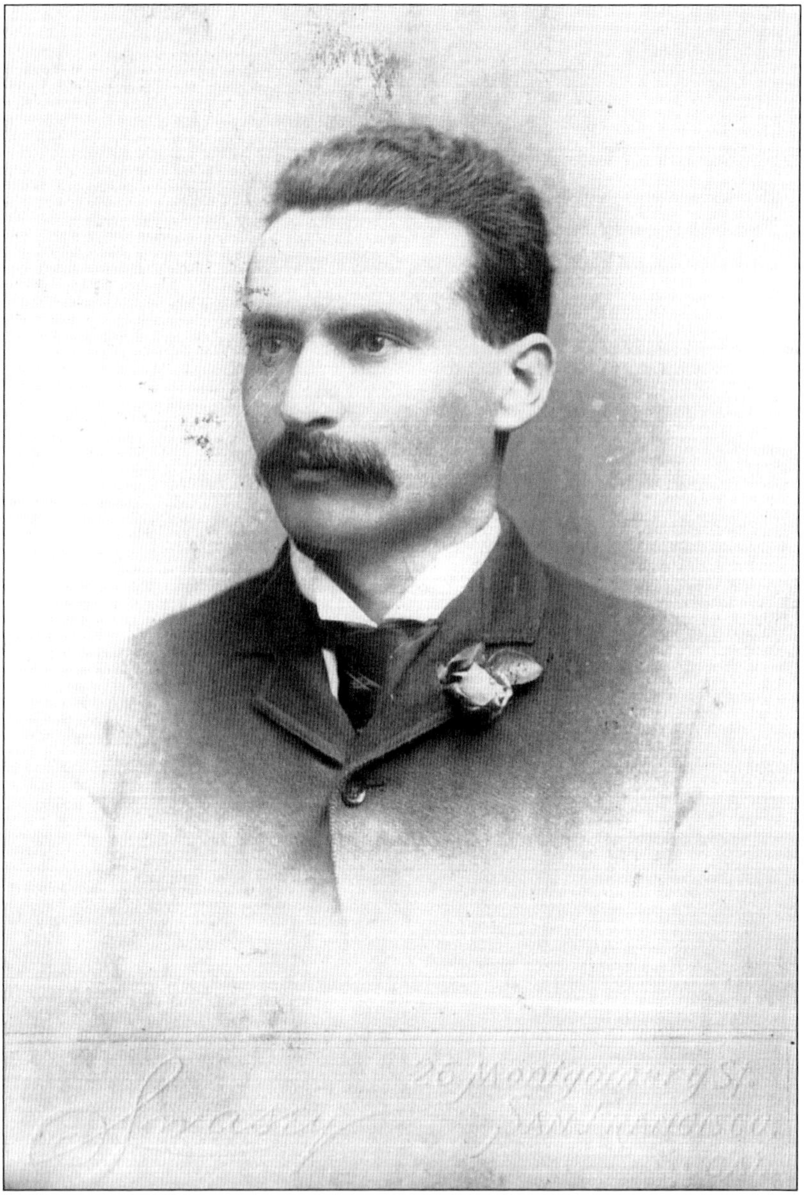

In 1890, a small group of San Francisco businessmen organized the Belvedere Land Company. They hired 26-year-old Michael O'Shaughnessy to survey their land and divide the lots. The Tamalpais Land and Water Company had also hired O'Shaughnessy to subdivide the former Throckmorton property, to find water for the property, and to sell off the subdivided lots. O'Shaughnessy's plan was carefully set out with quality in mind. The proposal included small clusters of commercial lots, large residential lots, and parcels reserved for parks, schools, and churches. The company opened its office at what is now 128 Throckmorton Avenue, in downtown Mill Valley. This work launched the career of O'Shaughnessy, who became one of the most well-known water engineers in the San Francisco Bay Area. Later, he was the city engineer for the City of San Francisco, developed the water system at Hetch Hetchy, and helped create the Marin Municipal Water District. (Courtesy of the Lucretia Little History Room, Mill Valley Public Library.)

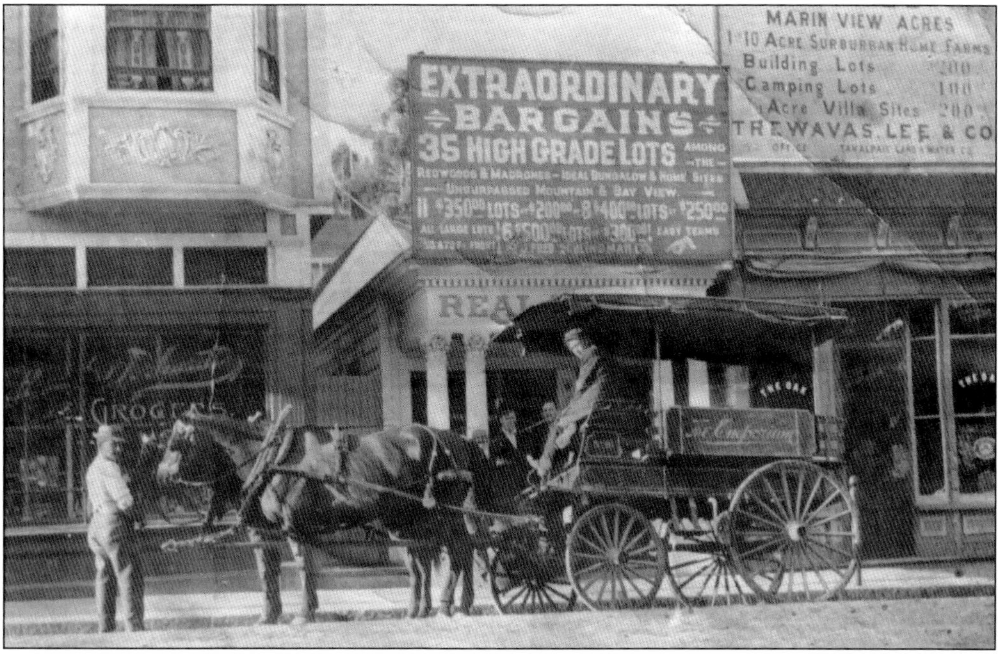

In 1890, more than 3,000 people attended the Tamalpais Land and Water Company land auction in Mill Valley. In excess of 200 parcels were sold, and on the same day many lots immediately sold again. The company created a land boom. This scene, from about 1910, showing the front of 118 Throckmorton Avenue in Mill Valley, shows the ongoing sale of subdivided lots 20 years after the land auction. (Courtesy of the Lucretia Little History Room, Mill Valley Public Library.)

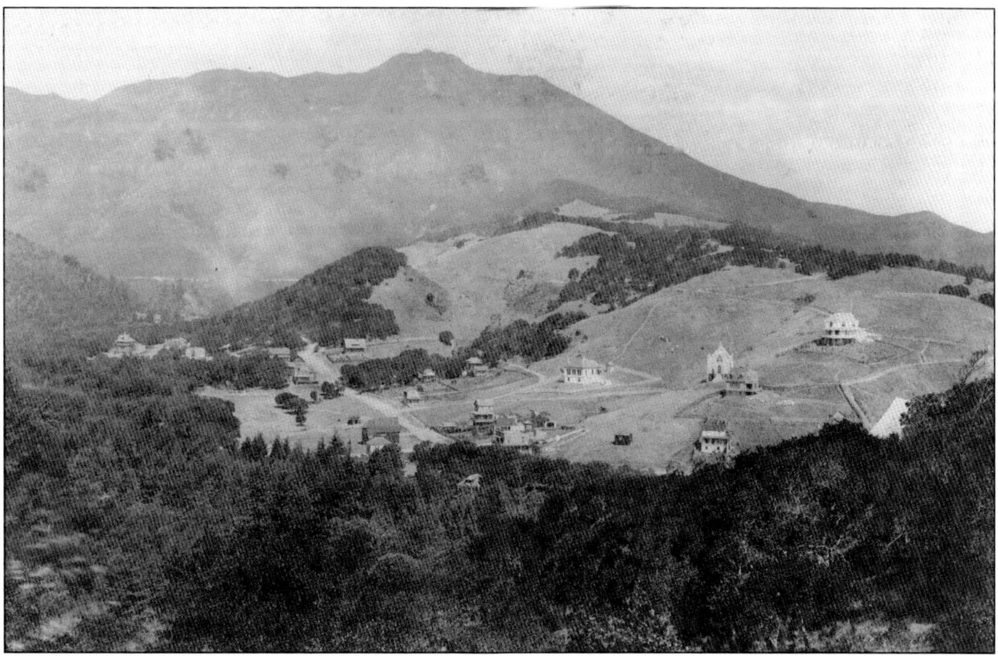

The Tamalpais Land and Water Company located at 128 Throckmorton Avenue in Mill Valley thrived after the land auction. This 1900 photograph shows the streets of Mill Valley as laid out by the company. (Courtesy of the Lucretia Little History Room, Mill Valley Public Library.)

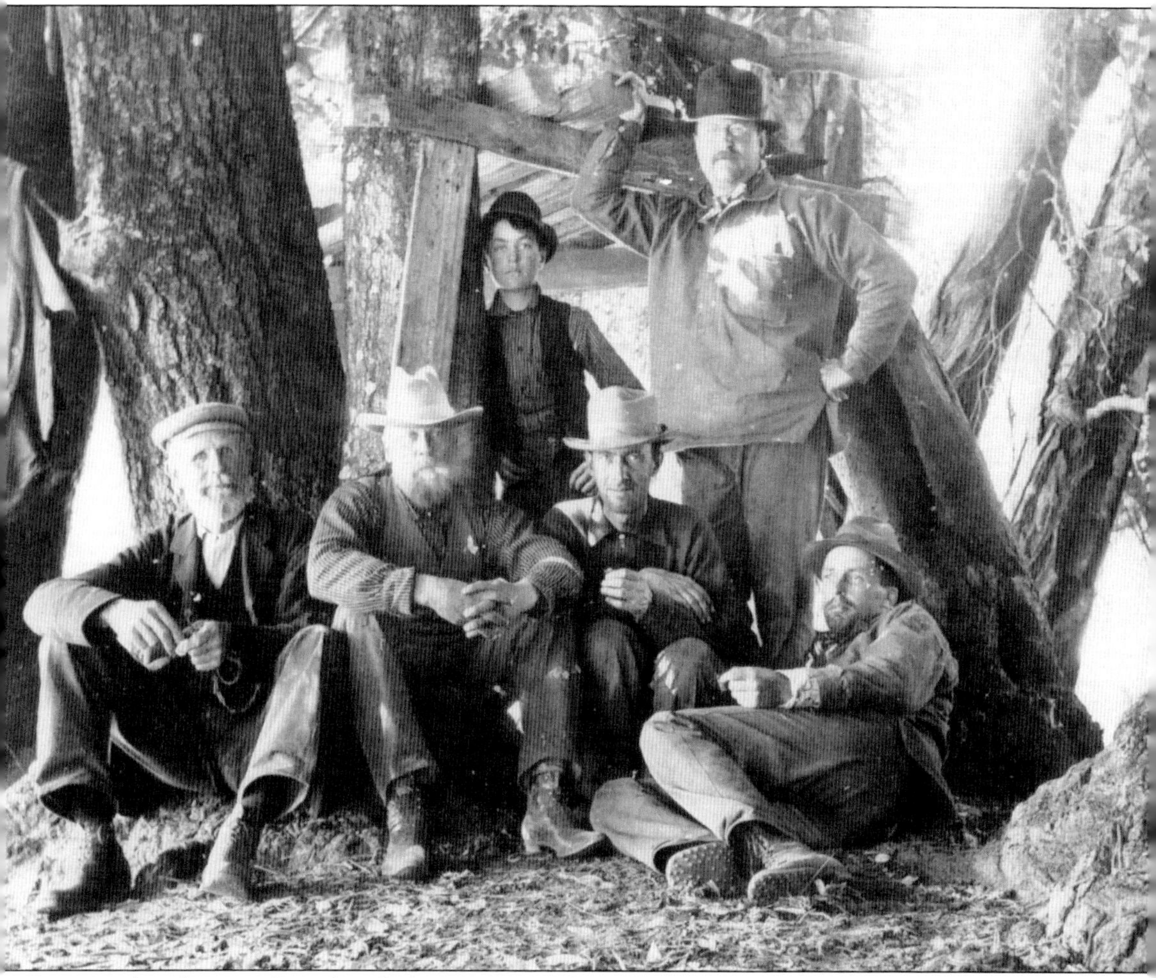

This photograph is identified as c. 1900 and includes, from left to right, John Muir (with cap); unidentified; John Eldridge Cushing, Sydney B. Cushing's son; William Kent (white hat, sitting); Sydney B. Cushing, Mount Tamalpais Scenic Railway partner (standing); and Eugene DeSalba, president of the Electric North Shore Railroad. Both the date and the identity of John Muir in this photograph are in doubt, for a few reasons. In 1905, when Kent donated Redwood Canyon (now Muir Woods) and offered to name it after John Muir, the two men had never met. Muir is known to have visited Muir Woods three times: once in 1904 without Kent, once in September 1908 at the invitation of Kent, and once in September 1909 for the completion of the Muir Woods Inn—too late for a photograph including Cushing, who died earlier that month. While Muir has been photographed hundreds of times, rarely—if ever—did he wear a cap as opposed to a more traditional hat, and by 1908, his beard appeared longer than shown in this image. (Courtesy of the Lucretia Little History Room, Mill Valley Public Library.)

Two

Publicly Owned Water

During the first decade of the 20th century, citizens in San Rafael, Ross Valley, and southern Marin organized a campaign to create a common water system. The laws of the day only allowed water districts to organize along existing county and town lines. With ready access to water from Mount Tamalpais, many citizens wanted a water district that would encompass a region beyond any town limit but within only a section of Marin County. New laws were needed to make that hope a reality. In 1910, newly elected Progressive governor Hiram Johnson signed legislation that finally made it possible to pass the Municipal Water District Act of 1911. In anticipation of the passage of that law, the citizens of Marin were prepared. They had been organizing for many months prior to the act's passage on December 23, 1911. In April 1912, with the direction of effective leaders and strong community support, the voters throughout San Rafael, Ross Valley, and southern Marin approved the creation of the Marin Municipal Water District (MMWD), making it the first municipal water district in the state.

As citizens organized and community leaders like Marin assemblyman George Harlan helped draft the necessary legislation, the powerful private water companies heartily campaigned against any change. The value of the private companies had to be established, which took months of contentious hearings before the California Railroad Commission, and bonds had to be issued and separately approved by the voters. The watershed lands and water delivery systems had to be purchased from the private water companies, using the power of eminent domain, and transferred into public hands. Congressman William Kent had long promoted the "park-like setting" of Mount Tamalpais, something the community gained as the mountain moved into public ownership. Immediately, this group of community leaders began preparing for the construction of Alpine Dam. The private companies continued to challenge the change in ownership, and it was another four years before MMWD began to deliver water.

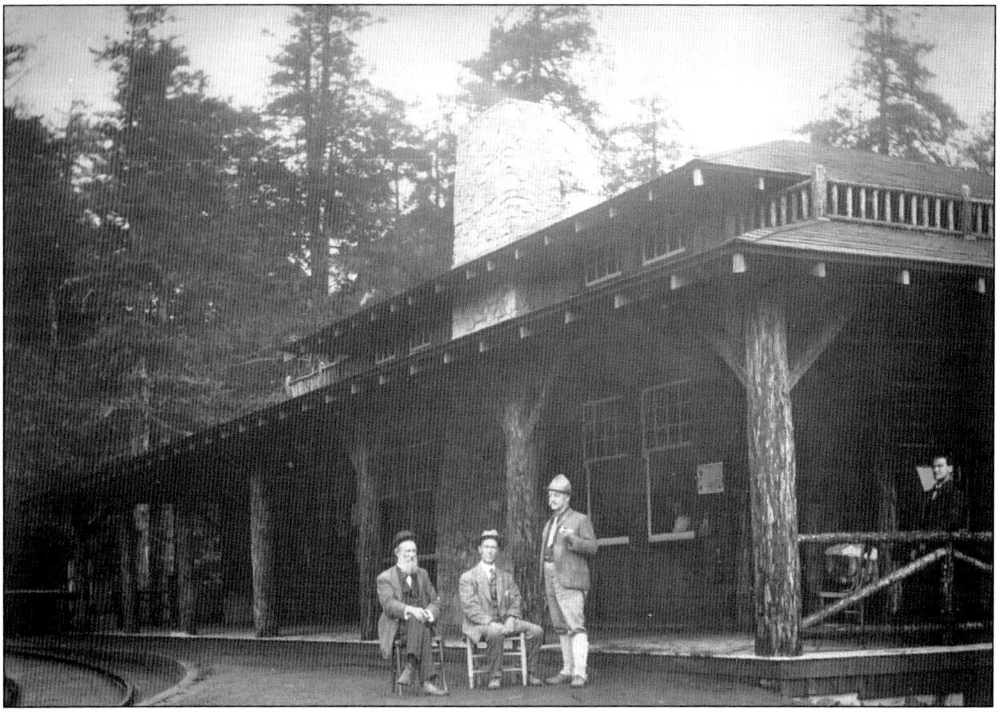

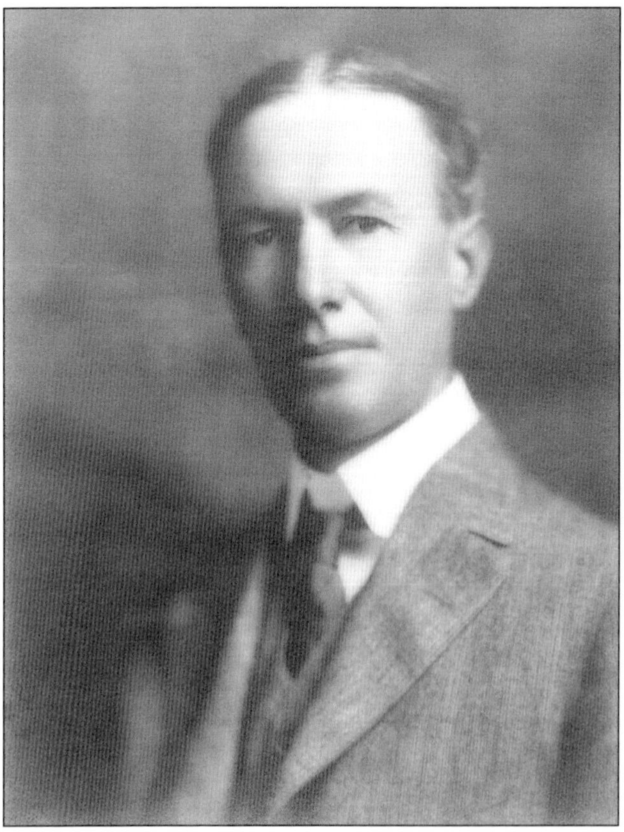

William Kent was raised in Chicago and spent summers in Marin. In Chicago, he was elected to serve as alderman and promoted urban reforms. In Marin, he promoted conservation and wilderness preservation. In order to save Redwood Canyon, Kent, pictured above in between John Muir (left) and Gifford Pinchot, head of the US Forest Service, purchased the canyon and donated it to the federal government as a park—Muir Woods National Monument. In 1907, when some people expressed concern that his gift deprived the community of a reservoir site needed for water supply, Kent accepted the challenge and spent the next 10 years promoting a public water system for Marin. (Above, courtesy of the Lucretia Little History Room, Mill Valley Public Library; left, courtesy of the National Park Service.)

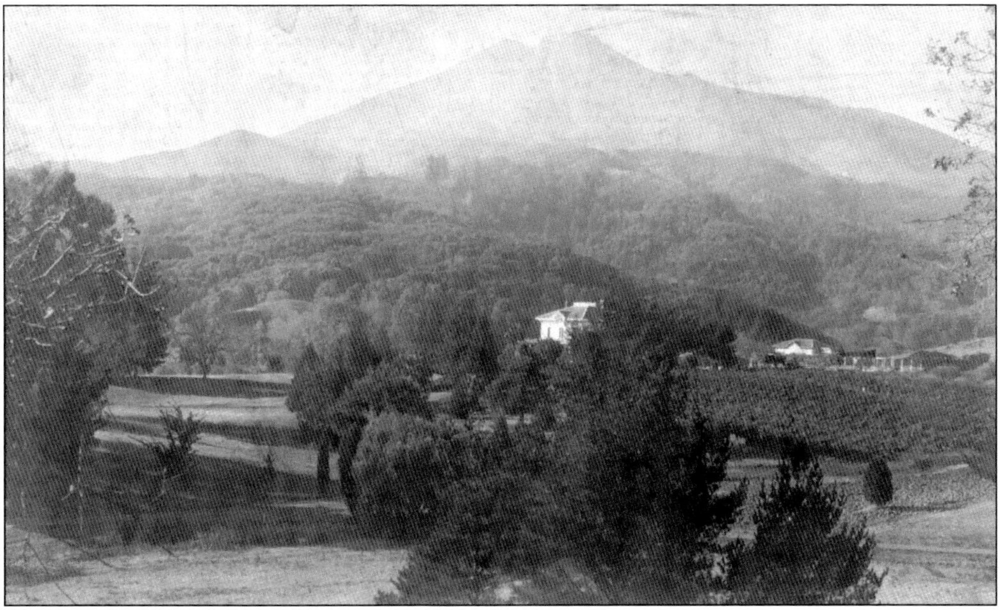

By 1900, the Kent family was one of the largest landholders in the county. Spending more and more time at his home in Marin (above and below), Kent took steps to make the county his permanent residence. In 1907, he announced his intention to run for county supervisor, only to find that he had not been a resident long enough to qualify for candidacy. He reset his goals and prepared to run for congress in 1910. The conservation movement was developing during those years, and it made for a good platform, especially against the background of the Muir Woods gift. (Photographs courtesy of the Anne T. Kent California Room, Marin County Free Library; below, contributed by J.R. Leavenworth.)

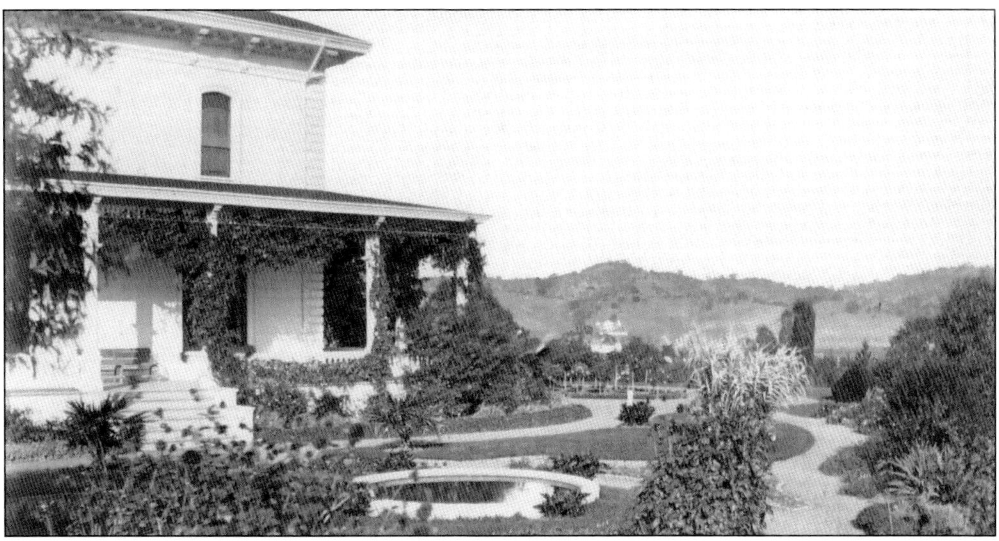

VOLUNTEERS

Carrying Initiatory Petitions

The following named persons have volunteered to carry petitions or have co-operated in initiating these proceedings taken for the purpose of organizing Marin Municipal Water District.

S. M. AUGUSTINE, San Rafael
TOBIAS HOCK, San Rafael
CLEMENT C. EDWARDS, San Rafael
L. E. FLEISCHMAN, San Rafael
C. S. WHITAKER, San Rafael
JAMES BEGLEY, San Rafael
WILLIAM L. COURTRIGHT, San Rafael.
S. K. HERZOG, San Rafael
S. J. NORTON, San Rafael
NEWTON E. CLEMENSON, San Anselmo
L. R. S. HOLMES, San Anselmo
G. N. VAN ORDEN, San Anselmo
JAMES A. SNOOK, Ross
JNO. W. GUNN, Kentfield
C. F. BURK, Kentfield
S. P. BROWNLEE, Kentfield
JONATHAN E. WEBB, Kentfield
E. L. MERWIN, Corte Madera
L. A. REA, Corte Madera
CASPER J. GARDNER, Mill Valley
N. P. YOST, Mill Valley
JACOB GARDNER, Mill Valley
W. G. MORROW, Sausalito
C. M. MOORE, Sausalito
G. H. SMITH, Sausalito
E. W. JACKSON, Sausalito

OBJECT OF THIS PETITION

To create a water district with in the area as outlined in the accompanying map, so as to enable the parties residing within said area to get the benefits of a municipally owned water plant.

HOW DOES IT AFFECT YOU, MR. SIGNER?

The expense of the water district, which include interest, depreciation and operating expenses, etc., will be paid out of the revenues of the water works.

That is what is being done now; all the revenue is derived from present water rates.

If you are a tax payer, but not a rate payer, the water district will not increase your taxes. It will render possible, however, if you should desire to lay out a tract of land or build a home, for you to get water supplied to such property by paying only the charges for service connections, so that you will not have to pay any more than does the man who at the present time has a pipe laid in front of his property who wishes to connect with same.

If you are a rate payer and a tax payer, you will save doubly. Your water rates will have a tendency to be reduced as a consumer, and as a tax payer you will be freed from the litigation expenditures which your City Government has been put to in the suits and differences with the Water Companies over rates, etc.

PRESENT CONDITIONS OF WATER SUPPLY

The present water works plants, sources of supply and additional sources of water supply are all owned by private corporations. It is the policy of one of these private corporations, to-wit: The Marin Water & Power Company, to require the property owner who wishes to lay a pipe in front of his property— i. e., a pipe extension—to pay the cost of the same. After said extension is made, the same becomes the property of the Water Company, thus making the property owner pay for the improvements to the Water Company's plant; and after paying for said improvements, if such property owner wants water, he is called upon to pay interest upon said improvements through water rates.

OUTLINE OF THE WATER DISTRICT PLAN

Method of Organization—

city or town at its (town or city) limits at a uniform rate to all within the district.

The district must sell water at cost, which shall include interest on bonds, sinking fund, maintenance, operation and deterioration charges.

Each Town Independent—

Each city or town is to own its own distributing system, collect its own water rates, etc., and thus each city will have its own municipal water plant, receiving water at its own town limits from a joint municipal plant (to-wit, the district).

Advantages—

The district plan enables the carrying out of a water supply project which is financially impossible for any one city within the proposed district. It will not impair the bonding capacity of any city, as all bonds will be raised by the district, except such bonds as each city will issue to buy the distribution system which lies within its limits.

Under present conditions the work of construction costs the private companies more than it will the district in two ways: The district can borrow money on bonds at a less rate than the companies can; also the district can dispose of its bonds at par, which the water companies are seldom able to do; as they are compelled to submit to heavy discounts and commissions.

ADVANTAGES WHICH THE PROPOSED WATER DISTRICT WILL BRING

It will enable the district to borrow money at a less rate of interest than a private corporation.

The bonds of the district will be sold at par, with no discounts or commissions charged in the sale of the same.

It will improve the water system by additions and enlargements.

It will lay larger pipes, thus giving more water and greater pressure.

It will place more fire hydrants, thus affording increased fire protection.

It will enable the community at large to share in the increase in the values of the properties of the water plant without being charged for said increased values in water rates. Increase of population means increase in value. Each additional inhabitant is an asset, as all require water.

As early as 1894, groups of Marin citizens began to hold meetings at "Y's Hall," in San Rafael, to express concern for the inadequate water system in Marin. Over the next 15 years, those concerns grew louder. By 1910, most areas of Marin developed "citizen water committees," and many were working collaboratively towards the goal of a publicly owned water system. The movement was popularly referred to as the "Free Water Movement." In 1911, the legislature passed laws that for the first time allowed for the creation of a public water system that could benefit all of Marin. The new legislation allowed for a water district that could cross town lines within a single county but not necessarily cover the entire county. Immediately, volunteers in Marin began to circulate petitions to gather the necessary signatures for voter approval of a publicly owned system. Many of the organizers listed on the petition were later involved in the development of MMWD, including Augustine, Whitaker, Webb, Gardner, and Morrow.

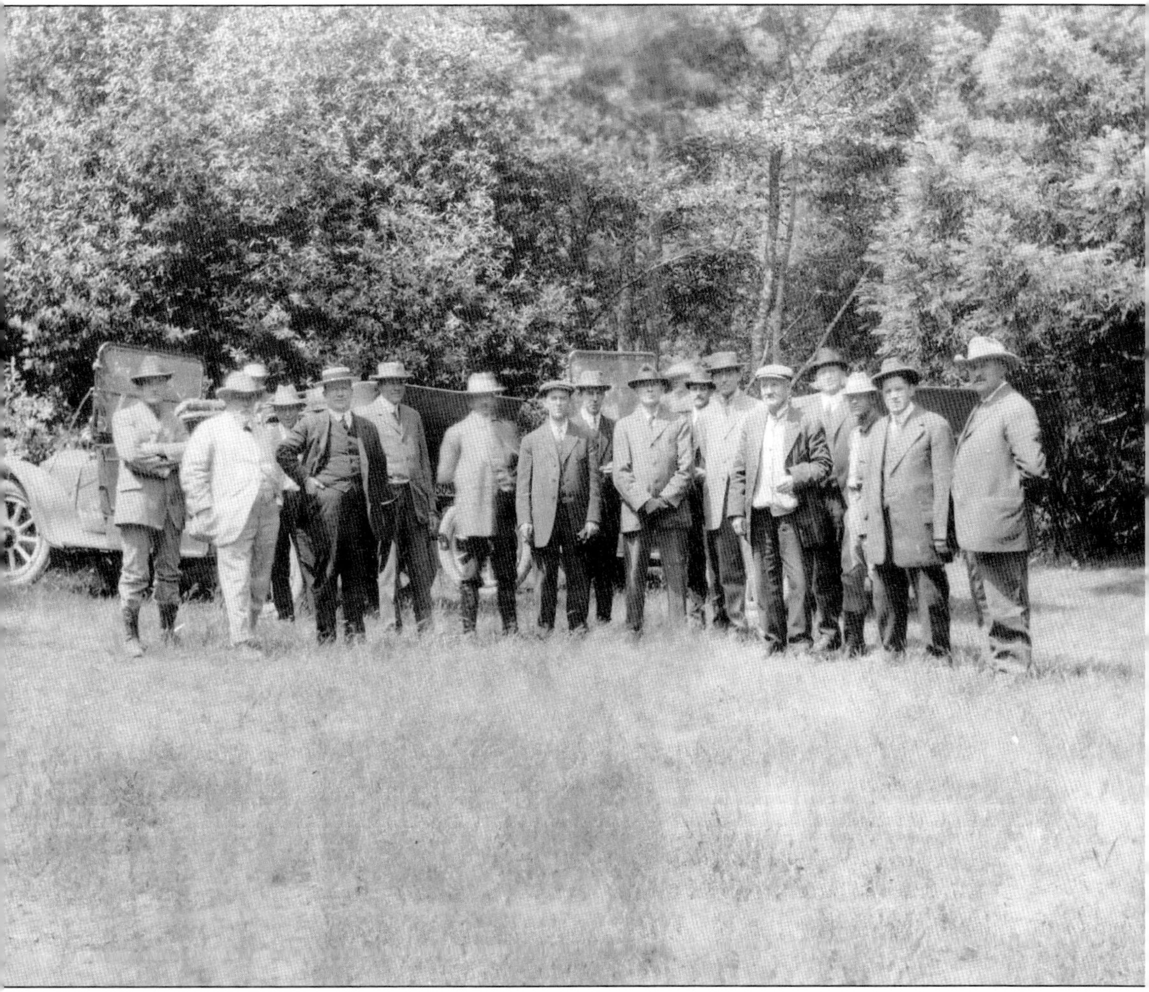

The above photograph, from about 1913, shows the first MMWD Board of Directors and other local dignitaries on a visit to the proposed site of Alpine Dam. In 1912, the board consisted of five directors elected by the voters and eight appointed directors—one from each of the incorporated towns in Marin—plus one appointed by the county supervisors. The first five board members to be elected were Michael O'Shaughnessy (Mill Valley), William Kent (Kentfield), Patrick Ross (San Rafael), William G. Morrow (Sausalito), and V.J.B. Cheda (San Rafael). In 1934, the state legislature revised the makeup of the board and eliminated the eight appointed positions. Today, the board is still made up of five elected members. Dr. Samuel M. Augustine (second from left, in the white suit) was the first manager of MMWD. He was an early Marin pioneer, and throughout his life had many community positions and duties, including medical doctor, superintendent of Marin schools, and chief cashier of A.P. Hotaling Bank in San Rafael.

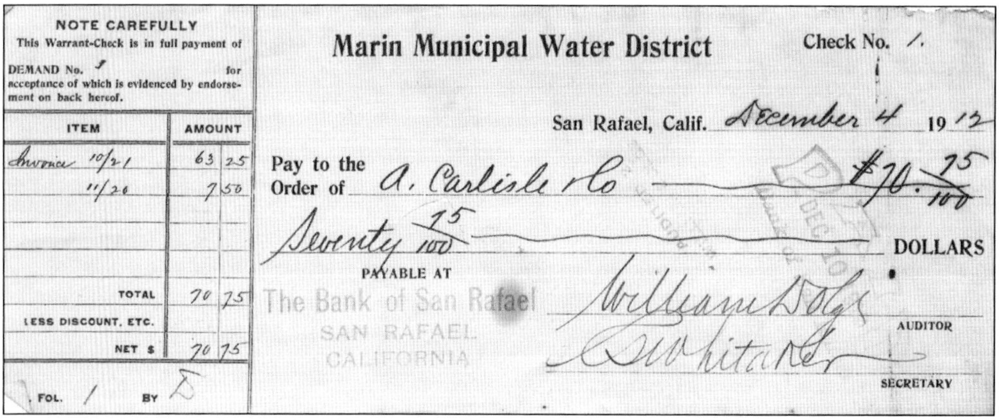

NOTE CAREFULLY
This Warrant-Check is in full payment of
DEMAND No. 5 for
acceptance of which is evidenced by endorsement on back hereof.

ITEM	AMOUNT
Invoice 10/21	63 25
11/20	7 50
TOTAL	70 75
LESS DISCOUNT, ETC.	
NET $	70 75

FOL. 1 BY D

Marin Municipal Water District
Check No. 1.
San Rafael, Calif. December 4 1912
Pay to the Order of A. Carlisle & Co $70.75/100
Seventy 75/100 DOLLARS
PAYABLE AT The Bank of San Rafael SAN RAFAEL CALIFORNIA
William Dolge AUDITOR
C S Whitaker SECRETARY

On August 5, 1912, MMWD held its first board meeting. That night, and in meetings over the next few months, the board appointed positions and established the organization of MMWD. It appointed Clarence S. Whitaker of San Rafael as secretary of the board, a part-time position. The board also hired William Dolge to serve as auditor, at a salary of $250. Together, Whitaker and Dolge set up a system of accounting for the district that included a banking relationship with the Bank of San Rafael, a successor of the A.P. Hotaling Bank. Samuel Augustine, who became MMWD's first general manager, was chief cashier of the bank. These two early checks were drawn on the bank by Whitaker and Dolge.

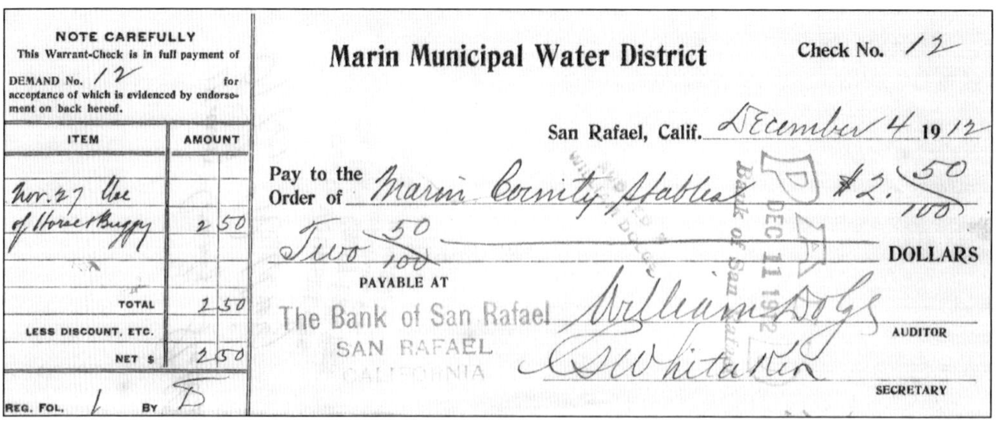

NOTE CAREFULLY
This Warrant-Check is in full payment of
DEMAND No. 12 for
acceptance of which is evidenced by endorsement on back hereof.

ITEM	AMOUNT
Nov. 27 Use of Horse & Buggy	2 50
TOTAL	2 50
LESS DISCOUNT, ETC.	
NET $	2 50

REG. FOL. 1 BY D

Marin Municipal Water District
Check No. 12
San Rafael, Calif. December 4 1912
Pay to the Order of Marin County Stables $2.50/100
Two 50/100 DOLLARS
PAYABLE AT The Bank of San Rafael SAN RAFAEL CALIFORNIA
William Dolge AUDITOR
C S Whitaker SECRETARY

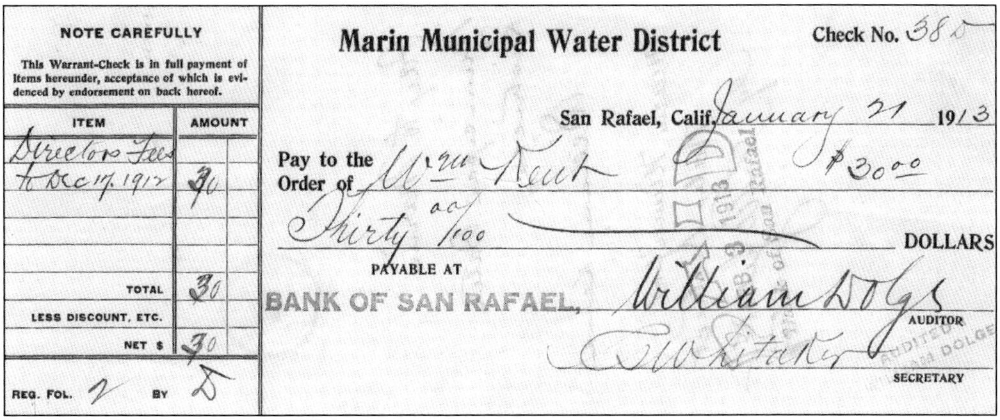

NOTE CAREFULLY

This Warrant-Check is in full payment of items hereunder, acceptance of which is evidenced by endorsement on back hereof.

ITEM	AMOUNT
Directors Fees to Dec 17. 1912	30
TOTAL	30
LESS DISCOUNT, ETC.	
NET $	30

REG. FOL. 2 BY D

Marin Municipal Water District

Check No. 385

San Rafael, Calif. January 21 1913

Pay to the Order of Wm Kent $30.00

Thirty 00/100 DOLLARS

PAYABLE AT BANK OF SAN RAFAEL, William Dolge AUDITOR

C W Whitaker SECRETARY

These two checks were drawn on the Bank of San Rafael for payment of directors' fees for board members William Kent and Michael O'Shaughnessy. Each board member received $30. At that time, Kent was serving in dual capacity, as a member of Congress from Marin and as a board member for MMWD. Understandably, he missed most of the MMWD board meetings because of his congressional duties and tendered his resignation to the board on February 18, 1913, having served for six months. Michael O'Shaughnessy actively served on the MMWD board until May 1916, serving as the head of the board's engineering committee during the planning for Alpine Dam.

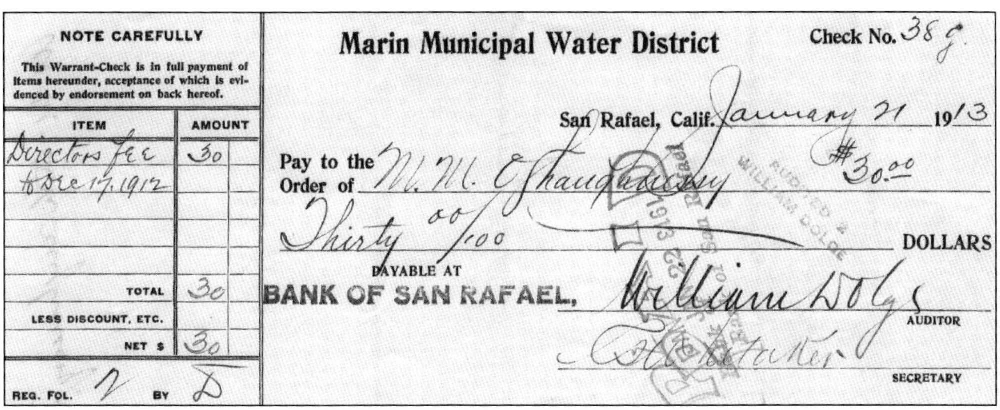

NOTE CAREFULLY

This Warrant-Check is in full payment of items hereunder, acceptance of which is evidenced by endorsement on back hereof.

ITEM	AMOUNT
Directors fee to Dec 17. 1912	30
TOTAL	30
LESS DISCOUNT, ETC.	
NET $	30

REG. FOL. 2 BY D

Marin Municipal Water District

Check No. 389

San Rafael, Calif. January 21 1913

Pay to the Order of M. M. O'Shaughnessy $30.00

Thirty 00/100 DOLLARS

PAYABLE AT BANK OF SAN RAFAEL, William Dolge AUDITOR

C W Whitaker SECRETARY

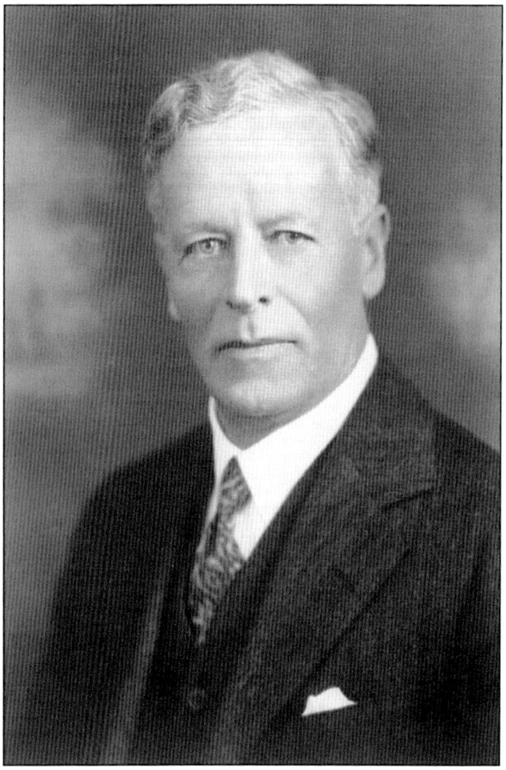

John Burt emigrated from Scotland to California in 1890 and was employed by the Tamalpais Land and Water Company to handle its waterworks. After working for its successor, the North Coast Water Company, he was appointed the first operating general manager of MMWD. He began his tenure in November 1916, when MMWD took over the other two water companies, and served until his retirement in 1935.

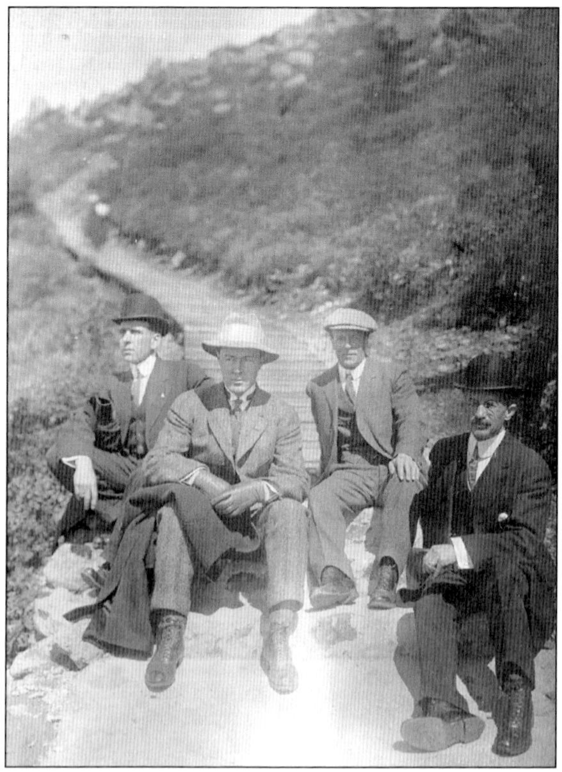

John Burt is pictured here on the Plank Walk Trail near the summit of Mount Tamalpais in about 1912. Burt, second from the left, is sitting with colleagues George Heller, Edwin Murray, and Horace Dorsey (in no particular order). Burt had worked in waterworks construction for seven years before coming to America from Scotland. (Photograph by F. Ransome; courtesy of the Lucretia Little History Room, Mill Valley Public Library.)

NOTICE!

To Water Companies and Owners Of Water Rights

Pursuant to a Resolution of the Board of Directors of the Marin Municipal Water District passed August 20th, 1912,

NOTICE

Is hereby given to all public or private corporations, firms or individuals who so desire to grant, transfer, lease with the privilege or purchasing or otherwise, or sell, any or all their water works or water works systems, waters, water rights, lands, rights and privileges, conduits, pipe lines, reservoirs, works, machinery, nad other property useful, or necessary to store, convey, supply, or otherwise make use of water for a waterworks plant or system for the benefit of said District, to deliver to the Secretary of said Board on or before

OCTOBER 15th, 1912

At eight o'clock P. M. at their headquarters in the Court House in the Supervisors' Chambers at San Rafael, Marin County, their proposal in writing and itemized.

Dated August 24th, 1912.

C. S. WHITAKER.

Secretary of the Board of Directors
Marin Municipal Water District.

The creation of MMWD was a grassroots community effort. When the first board of directors met, the members already knew each other and had worked well together for years. By the end of their second meeting, they had set up the district's organization, made key appointments, hired employees, and created the first water rate ordinance, Ordinance No. 1. Resolution No. 6 instructed its newest employee, board secretary Clarence S. Whitaker, to publish this notice to the community, indicating that those wishing to sell water storage systems, conveyance systems, or other elements of a water system should make a proposal to MMWD by October 15, 1912. Over the next few years, MMWD purchased the facilities of more than 26 separate water companies, ranging in size from the Marin Water & Power Company to those serving neighborhoods and even smaller ones serving only a few properties. From this assemblage of unlinked facilities, MMWD began the process of creating the integrated water system that serves the district today.

Directors

W. G. Morrow — W. E. Jones
V. J. B. Cheda — Dr. G. L. Bean
P. Ross — H. N. Thornton
M. M. O'Shaughnessy — N. B. Livermore
William Kent — E. W. Jackson
H. C. Symonds — C. C. Stevenson Jr.
R. H. Trumbull

Incorporated April 25, 1912

Marin Municipal Water District

Officers

R. H. Trumbull - President
S. M. Augustine - Gen'l Manager
A. R. Baker - Engineer
C. S. Whitaker - Secretary
William Dolge - Auditor
George H. Harlan - Attorney

Telephone S. R. 4501 — Court House

San Rafael, Cal. Nov. 12 1912

Mr. James S. Peters,
1414 Douglas St.,
Victoria.
B. C.

Dear Jim: I have a drafting job her for you at $110.00 per month, if you can take it on pretty short notice.

You said you would like to get down into this country, so now is your chance. The board here will cost you $30.00 a month, you will be 50 minutes from San Francisco and in a fine country and I am sure you will like it.

However, I can only promise you three or four months work but by that time there will be lots of work in the City and I am quite sure you can catch on again.

If you want to take a chance wire me at once, when you can leave.

Yours very truly,

ARB

On November 1, 1912, Albert Reed Baker was hired as the engineer for MMWD. Baker had extensive experience with water distribution systems, having worked in many parts of California and in Victoria, British Columbia. He had also worked for Michael O'Shaughnessy, who was by then the city engineer for San Francisco. With plans for Alpine Dam under way, and with the growing number of unrelated water systems being acquired by MMWD, a master plan was needed to develop a water distribution system. Over the next year, Baker worked to develop an integrated distribution system for the district. He reached out for experienced help, contacting James Peters of Victoria, British Columbia, whom he had experience working with. This letter shows Baker offering Peters a short-term job with MMWD. The abundance of drafting work in the aftermath of the 1906 earthquake in San Francisco allowed Baker to be encouraging. Peters first declined the offer but quickly recanted and took the "short-term" position with MMWD. Peters went on to become the general manager of MMWD and retired from the district 42 years later, in August 1954.

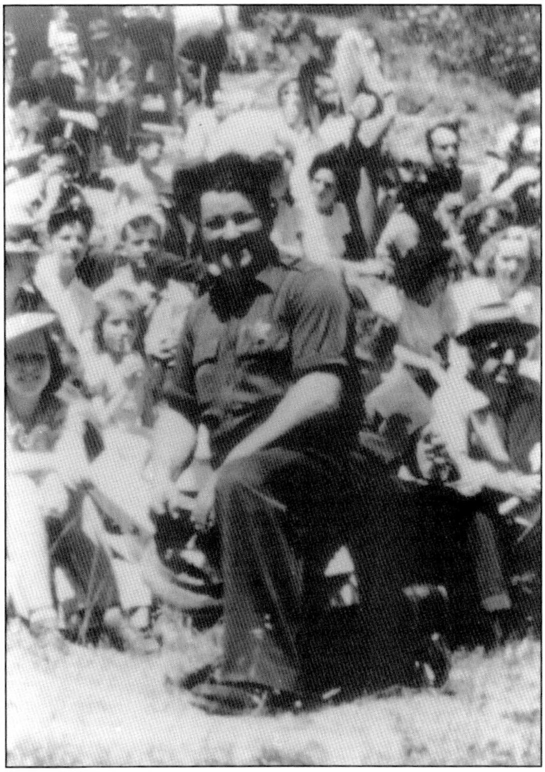

MMWD's park ranger program was formed in 1917, making it one of the oldest such programs in California. Among the communications to the board in 1917, district staff recommended that up to 12 patrolmen be hired to take charge of fire patrol, fire fighting, and enforcing the rules and regulations of the district. It also stipulated that the head patrolman furnish his own horse. Patrolman Joseph "Joe" Zapella, shown in both photographs, worked for the district for 27 years, from 1941 until his death in 1968. He was often called the "Old Man of the Mountain," not because of his age but because he had worked on Mount Tamalpais for so long.

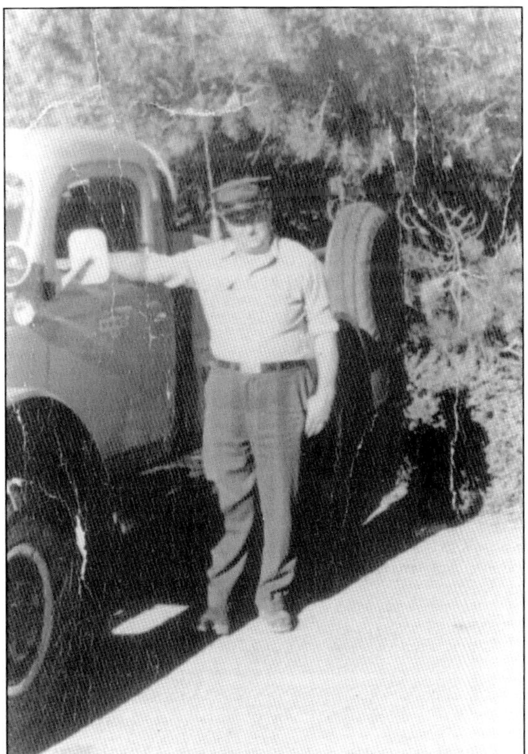

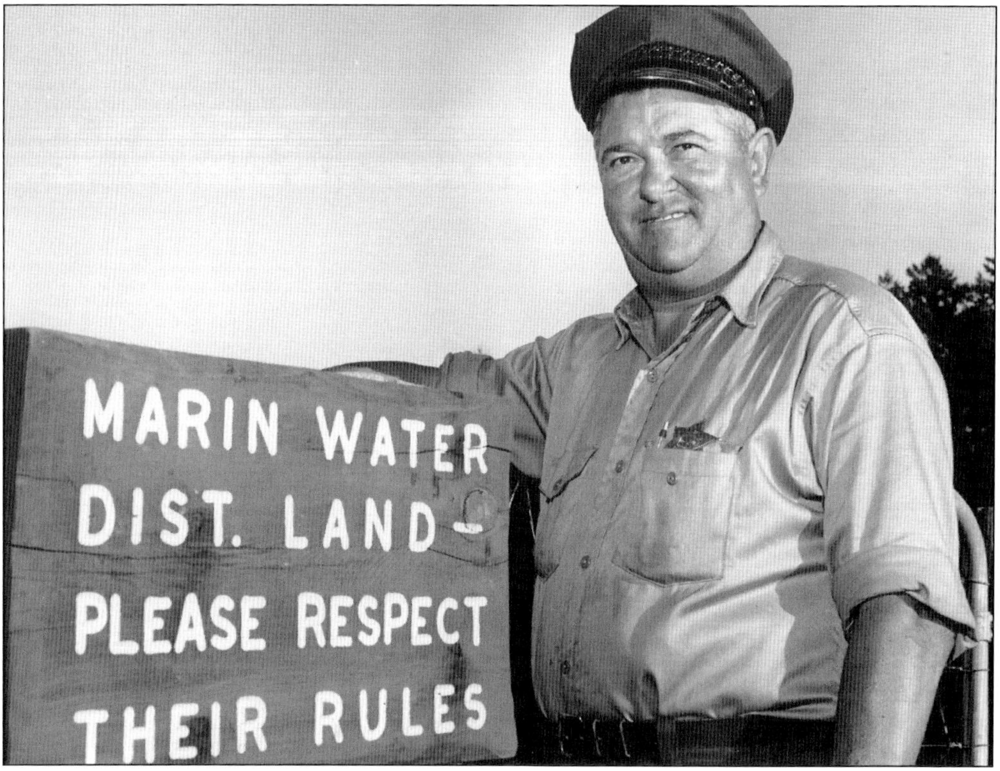

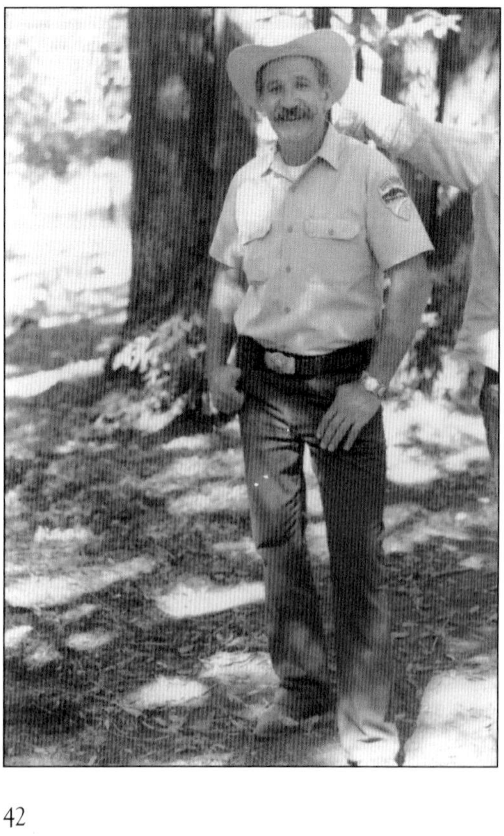

The Marin Municipal Water District has a long tradition of hiring people who end up spending their entire careers in the service of the district. Patrolman Joe Zapella (above) was featured in a *Marin Independent Journal* interview published in 1962 and there is a memorial in his honor on Mount Tamalpais that reads: "Joseph Zapella, Friend of the Hiker." In November 1946, Melvin J. "Moose" Muzinich (left) was hired as a patrolman and relief lake keeper and then reclassified as a ranger in 1975 in keeping with changes to the rangers' authority. He worked for MMWD for a little over 40 years, all the way up until 1987.

Three

WATER GETS TO THE COMMUNITY

After the 1906 earthquake, Marin's population grew rapidly as people fled San Francisco. The already severe water problem in southern Marin became even more pressing. The Marin Water & Power Company had constructed the Phoenix Gulch Dam in Ross in 1905, but it was not enough.

In August 1912, the MMWD Board of Directors held its first meeting and quickly moved to the planning stages. They appointed an engineering committee with O'Shaughnessy as chairman. They hired the district's first engineer, Albert R. Baker. Baker assessed the water supply and presented two options: bring water from Lagunitas Creek by a gravity system to a new reservoir and dam, or bring water from the Russian River in Sonoma County. With the desire to keep Marin independent, the board decided to construct the more local gravity system and divert water from Lagunitas Creek to a new reservoir created by the construction of Alpine Dam.

In August 1915, three years after the first water district board meeting, the voters of the district approved a $3-million bond issue. In their attempt to stop the public water movement, the private water companies vigorously opposed the bond campaign.

MMWD's board understood that without funding the water district would be nothing more than a paper organization. After approval and sale of the bonds, the water district had the funds to purchase the assets of the private water companies, to build Alpine Dam, and to buy the large Howard Shafter Lands for additional watershed.

A contract to construct Alpine Dam was awarded on June 2, 1917. As construction continued, the outbreak of World War I caused difficulties for most of the contractors. Machinery and construction materials became difficult or impossible to obtain. Contractors had made their bids at the beginning of the war when labor costs were 30¢ per hour, but they rose to 50¢ per hour by 1918. In January 1918, the water district and the bonding companies took over completion of the project, and the dam was completed on January 7, 1919.

Marin Water & Power Company

NOTICE TO CONSUMERS

San Rafael, July 27, 1916.

Supplementing our notice to you of May 16th last, we regret being obliged now to notify you that from and after August 1st the Company will be unable to furnish water for other than domestic purposes, and that all irrigation must be discontinued until further notice.

A hearty compliance with this notice is most earnestly and respectfully requested.

Marin water supply relies on local rainfall, received mostly from December through March. Over the last 130 years, rainfall at Lake Lagunitas has averaged approximately 52 inches per year, but with a wide range of high and low levels. Lake Lagunitas received over 100 inches of rain in 1890 and in 1895, but in 1929, 1931, 1939, 1947, 1976, and 1977, the lake received less than 30 inches of rain. Growing populations and an insufficient number of reservoirs to store the seasonal rainfall also caused periodic water shortages. Despite receiving more than 60 inches of rain in 1915, the shortage that generated this 1916 Marin Water & Power Company notice was mostly the result of the post-earthquake population boom in Marin and demonstrated to the county the need for more reservoir storage, ultimately provided by the construction of Alpine Dam.

In 1915, three years after MMWD was formed, those opposed to a publicly owned water district were still actively working against it, as seen in these election circulars. MMWD proposed its first bond issue for $3 million in 1915, the funds required to build Alpine Dam and purchase the assets of the private water companies—including 5,500 acres of Mount Tamalpais watershed lands. On August 28, 1915, the voters approved the MMWD request, which received a favorable vote from 77 percent of the district's voters. The "Vote No!" poster was an attempt by the Marin Water & Power Company to generate support for its proposal to build Tamalpais Dam as an alternative to Alpine Dam. The Tamalpais Dam, if it had been constructed, would have been located in the same general vicinity of Alpine Dam.

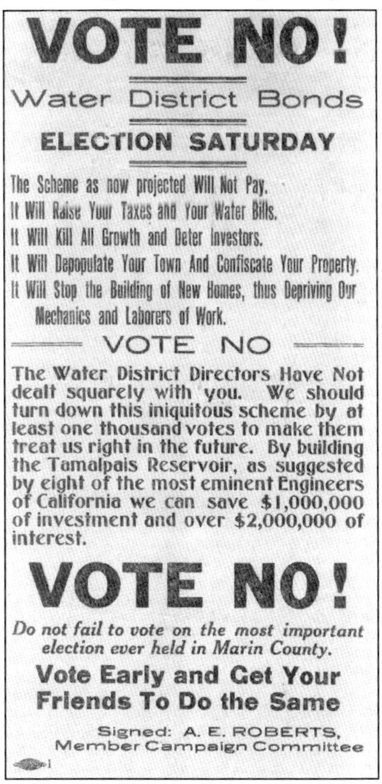

VOTE NO!

Water District Bonds

ELECTION SATURDAY

The Scheme as now projected Will Not Pay.
It Will Raise Your Taxes and Your Water Bills.
It Will Kill All Growth and Deter Investors.
It Will Depopulate Your Town And Confiscate Your Property.
It Will Stop the Building of New Homes, thus Depriving Our Mechanics and Laborers of Work.

VOTE NO

The Water District Directors Have Not dealt squarely with you. We should turn down this iniquitous scheme by at least one thousand votes to make them treat us right in the future. By building the Tamalpais Reservoir, as suggested by eight of the most eminent Engineers of California we can save $1,000,000 of investment and over $2,000,000 of interest.

VOTE NO!

Do not fail to vote on the most important election ever held in Marin County.

Vote Early and Get Your Friends To Do the Same

Signed: A. E. ROBERTS,
Member Campaign Committee

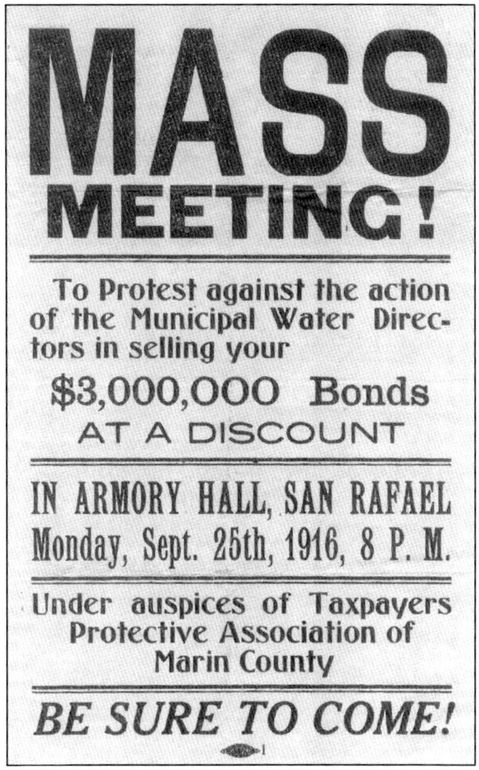

MASS MEETING!

To Protest against the action of the Municipal Water Directors in selling your

$3,000,000 Bonds
AT A DISCOUNT

IN ARMORY HALL, SAN RAFAEL
Monday, Sept. 25th, 1916, 8 P. M.

Under auspices of Taxpayers Protective Association of Marin County

BE SURE TO COME!

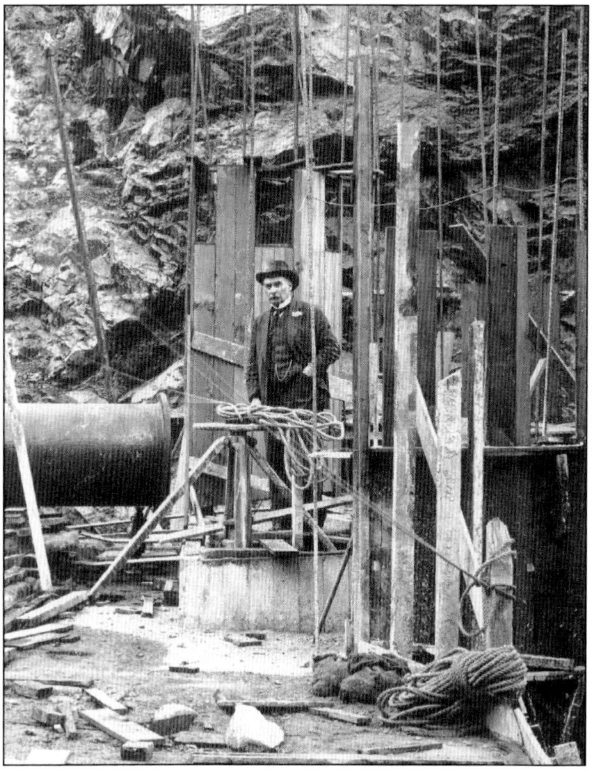

Michael O'Shaughnessy, shown on the Alpine Dam worksite in 1918, was the engineer in charge of the construction. He developed a good standing in the community working for the Tamalpais Land and Water Company in the 1890s. He worked as an advisor throughout the early years of MMWD and was elected to serve on the first MMWD Board of Directors.

O'Shaughnessy, shown below with his five daughters, had influence reaching beyond Marin County. In 1912, he was appointed by San Francisco mayor James Rolph as that city's chief engineer. He oversaw the rebuilding of San Francisco after the earthquake, including the development of the Hetch Hetchy water system, which began bringing water from Yosemite to San Francisco in 1934, only 16 days after O'Shaughnessy's death.

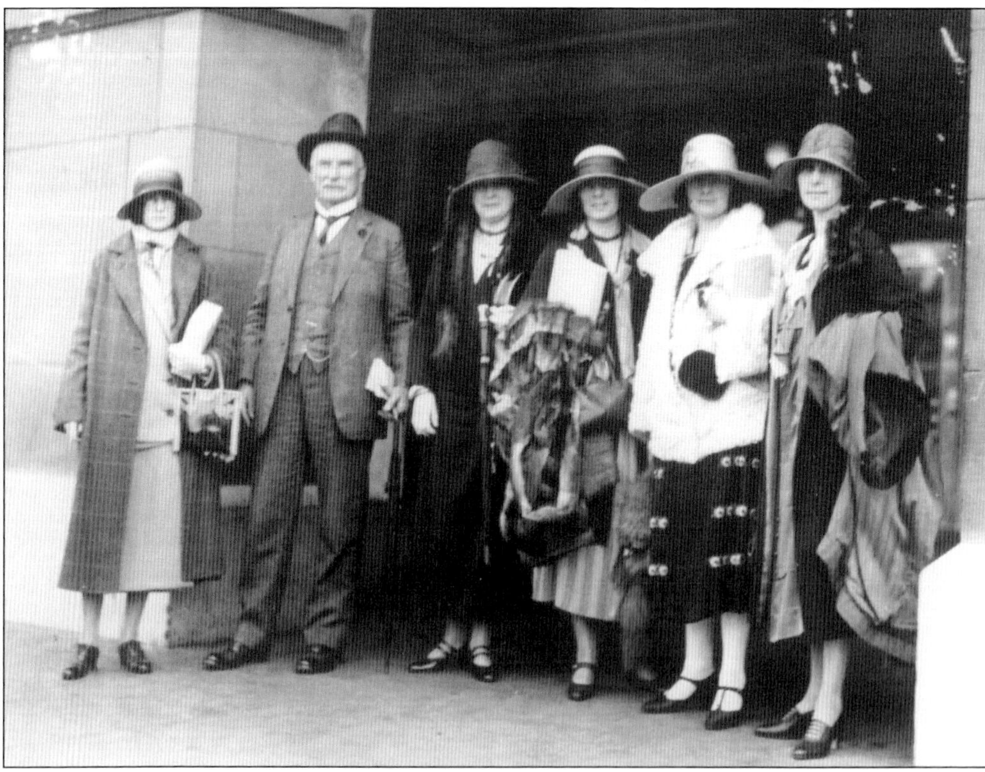

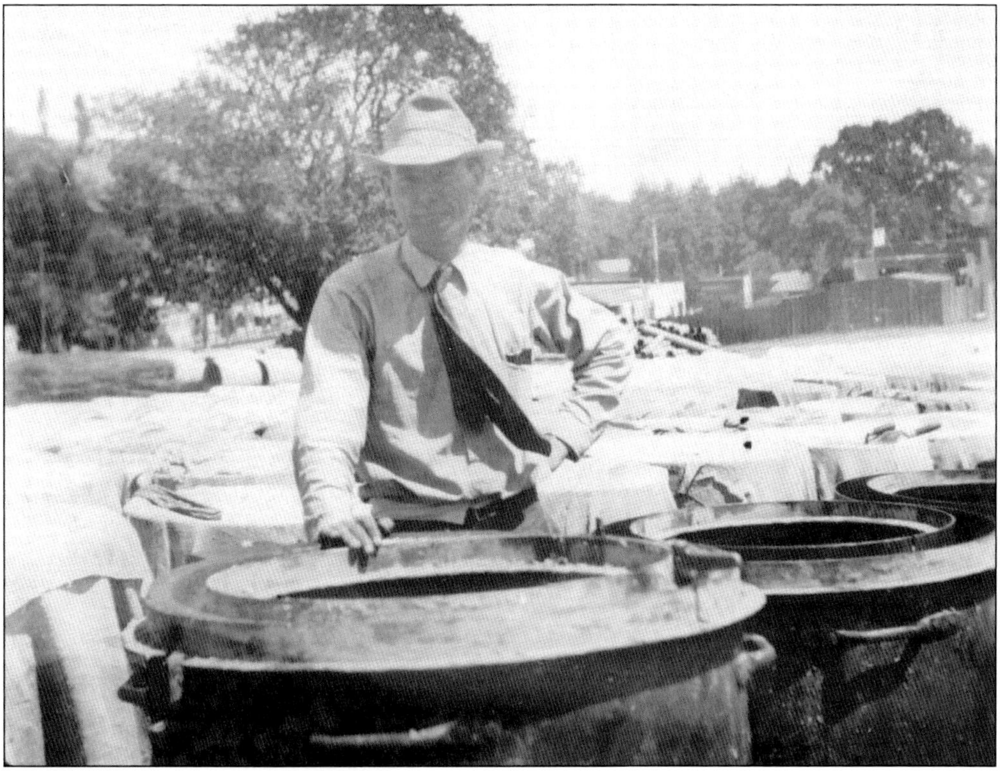

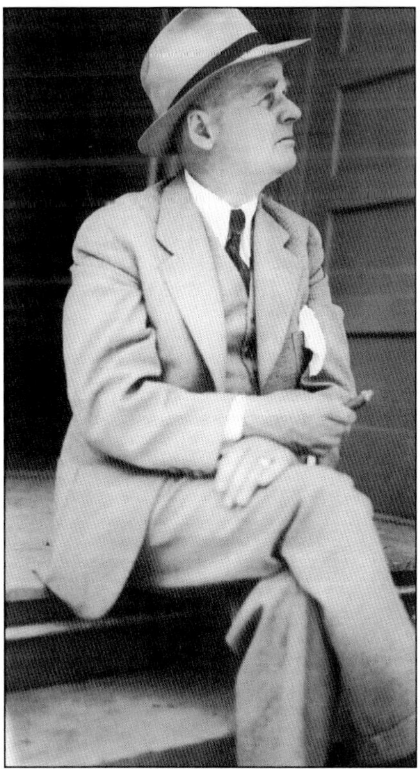

Albert Reed Baker (above) was born and raised in San Diego. He graduated from the University of California as a civil engineer in 1905. On November 1, 1912, he was employed as the engineer for MMWD and asked to investigate available sources of water for the district, including water from the Russian River. His recommendation, the Alpine Gravity System, is still used today.

Baker recruited James Peters (left) to work for MMWD, and he was hired in 1913 to work as a draftsman developing Alpine Dam. In 1933, he became the first assistant general manager of the district, and from 1935 to 1954 he served as the general manager. He was instrumental in developing the current organization of MMWD.

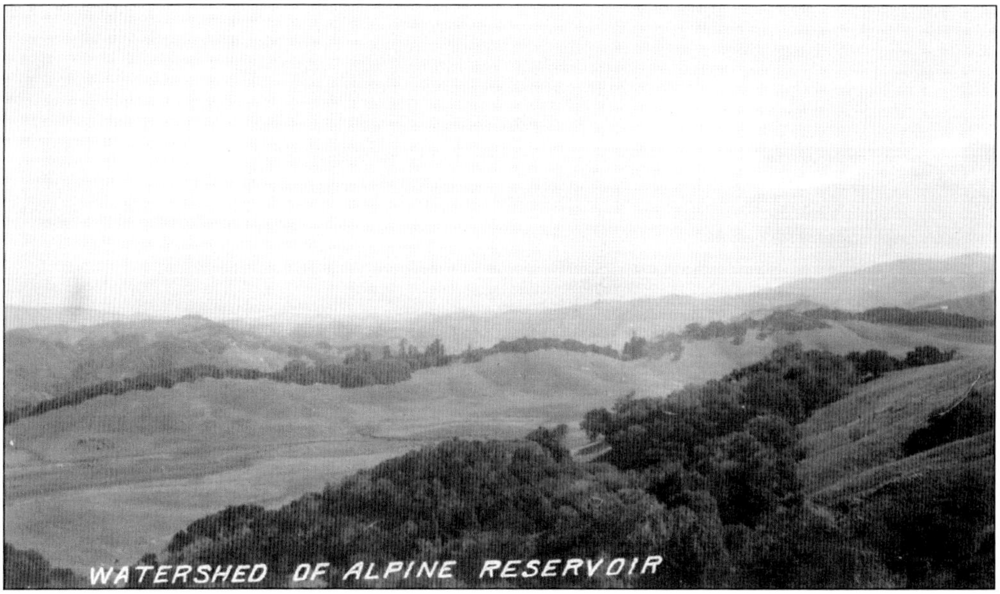

These 1917 photographs show the area of Lagunitas Canyon that was flooded two years later by the creation of Alpine Dam. The Lagunitas Canyon, located approximately seven miles from Fairfax, was the site for the dam. Plans for Alpine Dam and its resulting reservoir submerged several sections of the Bolinas-Fairfax Road, a county road, so it had to be closed and rerouted to make way for the project. Lagunitas Creek is to the right of the dirt road seen in the below photograph and drains 103 square miles of west-central Marin County, the largest watershed in the county. The creek originates on Mount Tamalpais and flows 22 miles before emptying into the southern end of Tomales Bay.

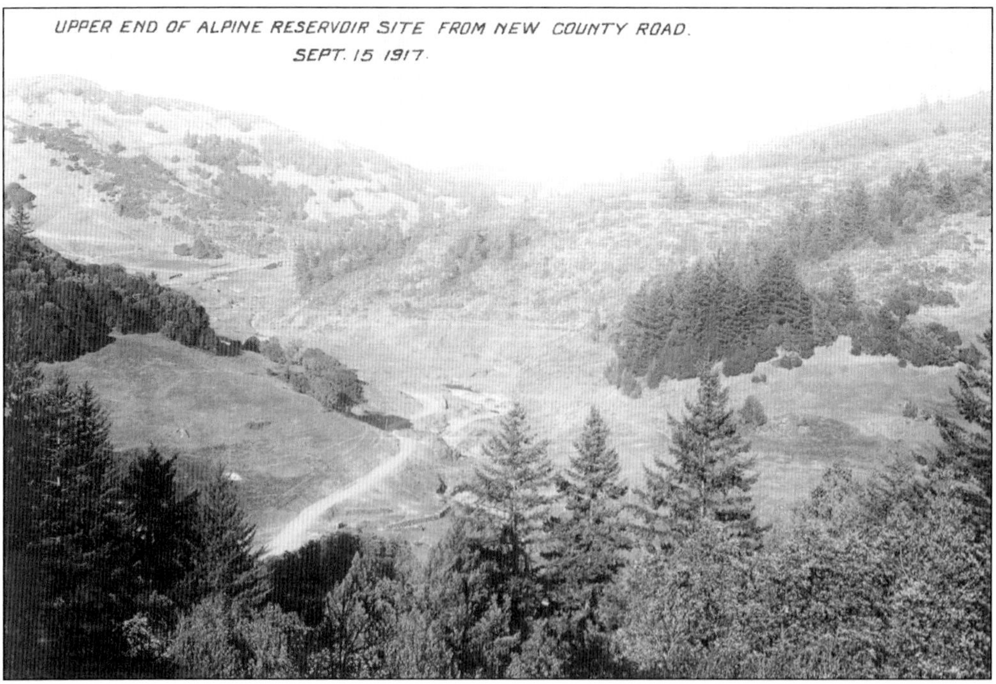

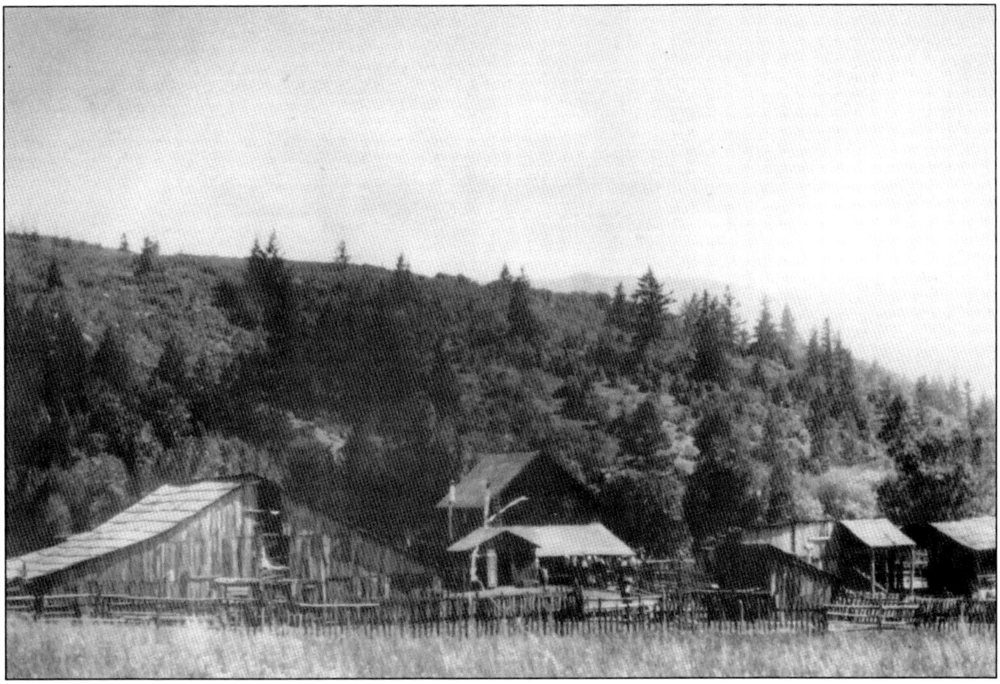

The Liberty Ranch property (above), now submerged below Alpine Lake, was damaged by fire in 1882 and again in 1885. In 1891, Vincent Liberty left the property for the East Bay, but Mrs. Liberty remained and attempted to run the property as a ranch and resort. In 1892, the property was again struck by fire and in February 1894, she gave up, sold the property to a Mr. Barfoldi, and moved to San Rafael. When MMWD was building Alpine Dam in 1918, the board of directors was asked to change the name of the property back to Liberty. The request came out of patriotic sentiment against the backdrop of World War I, not because of its former owners. (Above photograph courtesy of the Lucretia Little History Room, Mill Valley Public Library.)

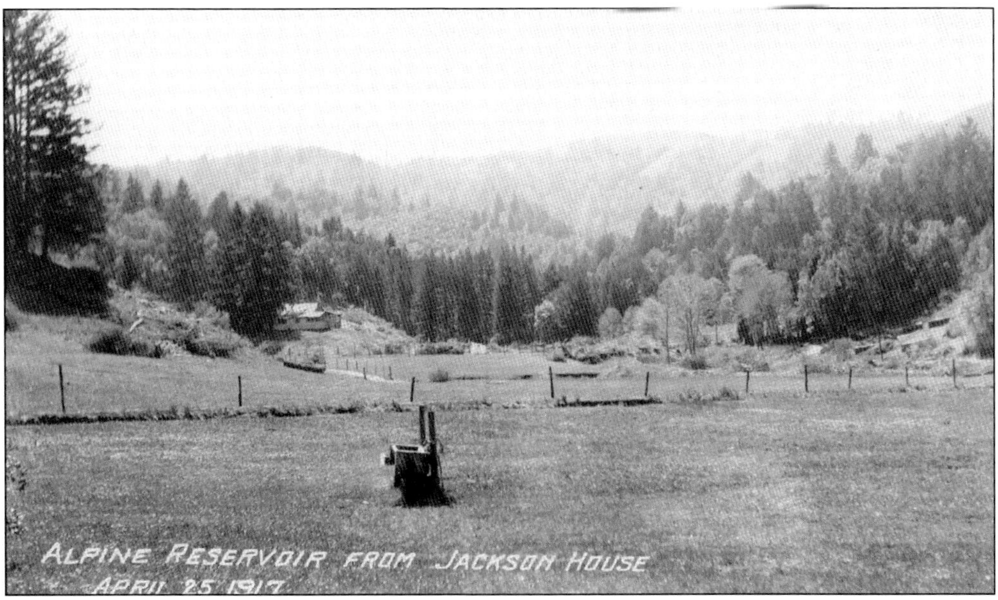

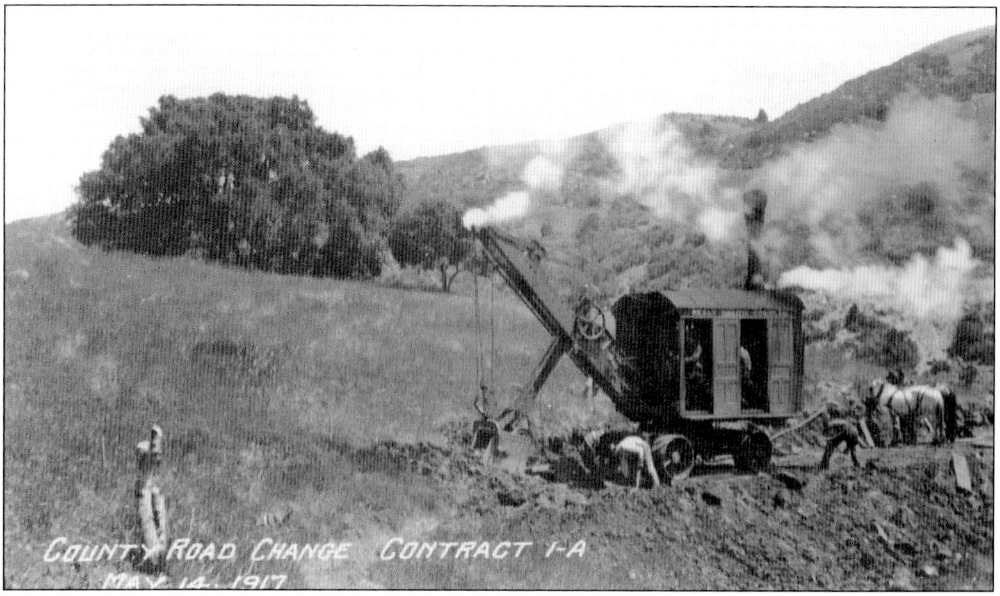

These May 1917 photographs show work being performed on the relocation of portions of the old Bolinas-Fairfax Road. In February 1917, the board of directors entertained seven bids for the road relocation. Under Contract 1-A, the job was awarded to A.J. Fairbanks of Willits to build 4.5 miles of road for $22,760.78. This belching steam shovel was the heaviest equipment used on the job. That period of time marked the transition between 19th-century reliance on horse-drawn labor and the development of motorized heavy equipment. Below, work is being done at the connection of the old Bolinas-Fairfax Road and the new road.

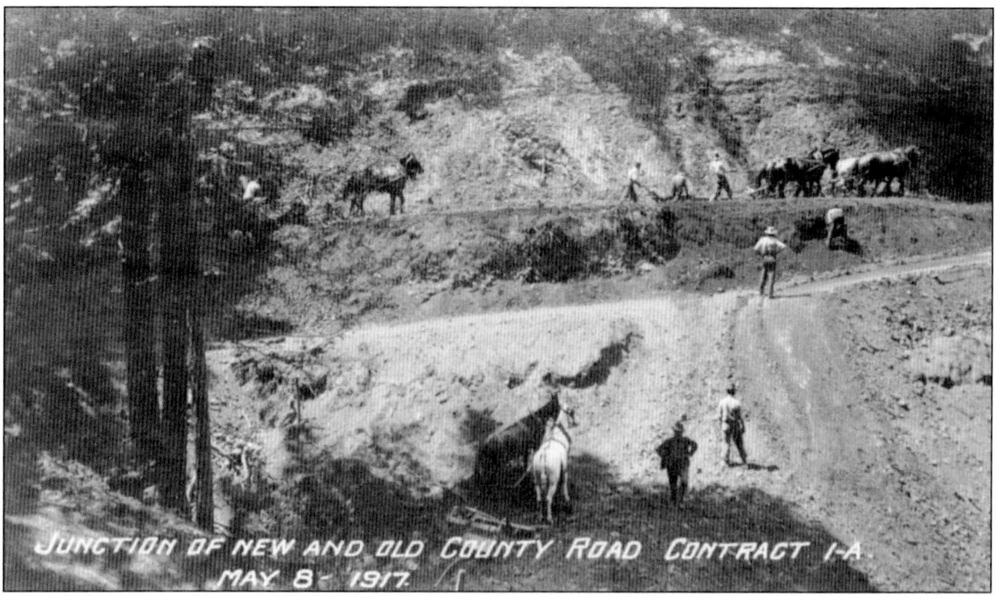

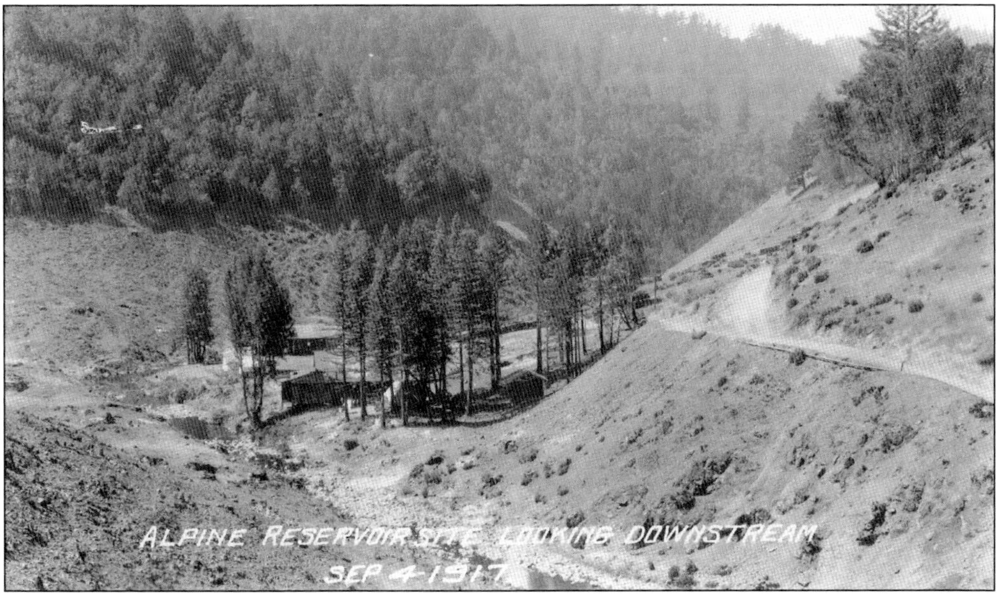

Much of the design work for the Alpine Dam and Reservoir project was done during the fall of 1915 and the spring of 1916. By October 1915, most of the tenants living on the valley floor had moved out, and Contract 1-B was given to S.P. Brownlee of San Rafael for the work of stripping—clearing and grubbing—the site. By the end of that month, the valley that would become Alpine Reservoir was cleared of rock, stumps, logs, and rubbish. When the lake level is low, foundations and chimneys from condemned houses and ranches in the valley can still be seen.

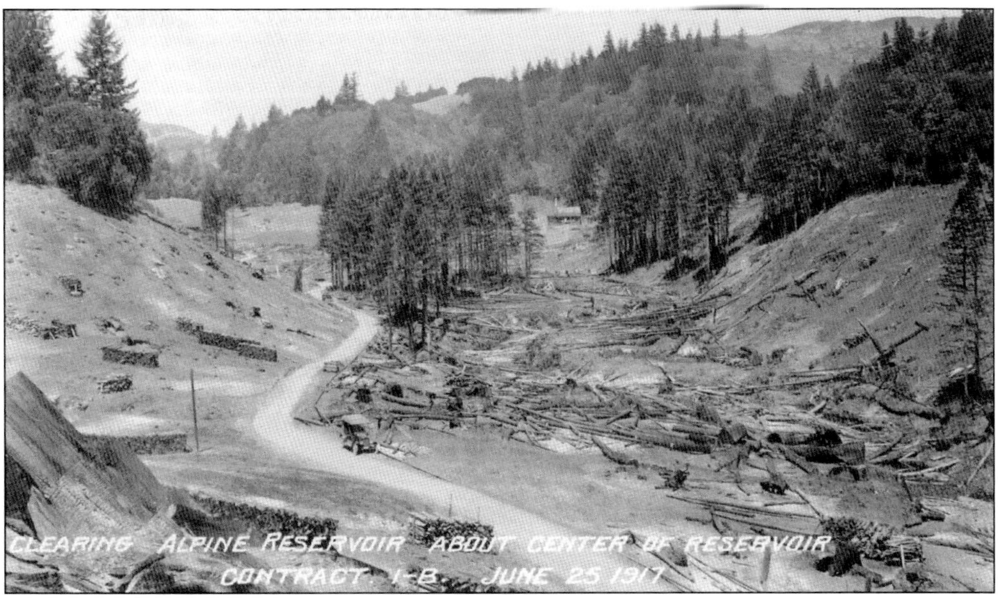

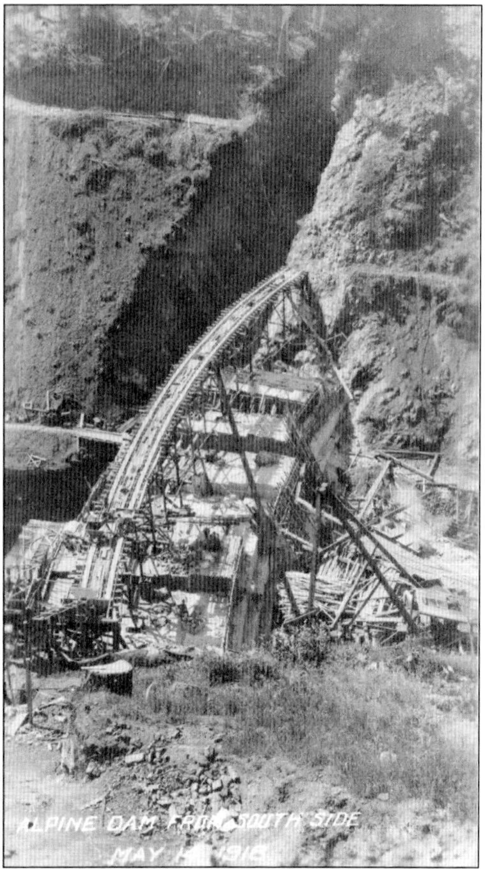

The contract for construction of Alpine Dam (Contract 2-C) was awarded to Leonard and Hansen on June 9, 1917, for their bid of $211,626.25. Planning for a dam at that site had been in progress for years, and preliminary construction work began one month after the contract was awarded. By August, equipment was on the site, and in October, a quarry was opened and concrete was poured. But with the advent of World War I and necessary resources for the war effort, the contractors were unable to obtain materials or meet their rising labor costs.

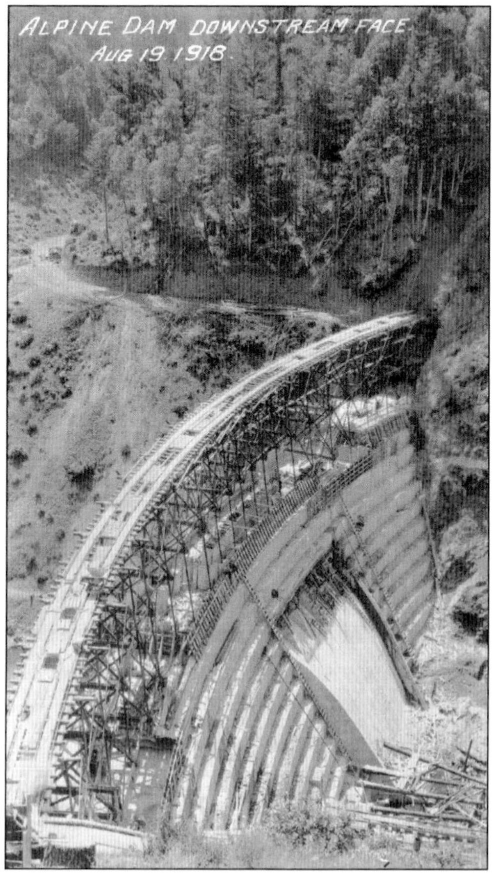

Because of the shortage of materials and climbing labor costs caused by World War I, the contractors were not able to complete Alpine Dam. In January 1918, MMWD was forced to take over the project and finish construction. The images here and on the preceding page show the stages of construction throughout 1918, in May, August, October, and December, respectively. The dam was completed on January 7, 1919. When completed, it was 100 feet high, 325 feet along the crest, and 15 feet wide along the top—just wide enough for the reopened Bolinas-Fairfax Road. The dam created Alpine Lake, which covered 135 acres and stored 1 billion gallons of water, an increase of 300 percent over the volume held at the combined Lagunitas and Phoenix Lakes.

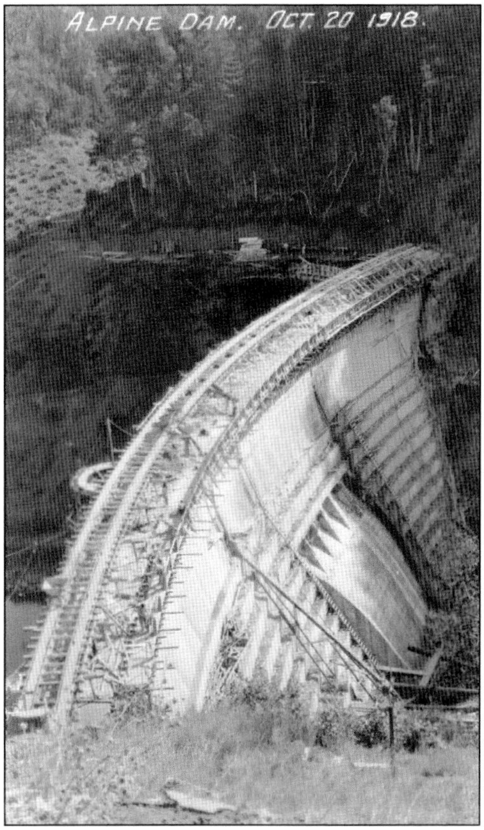

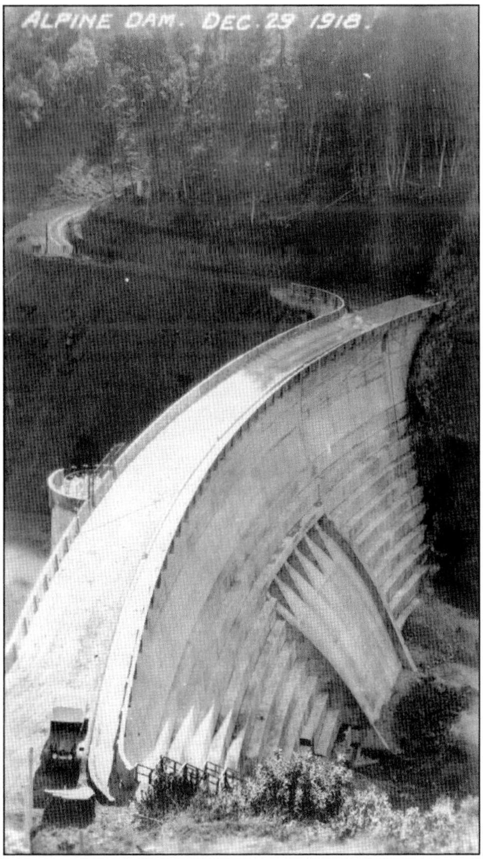

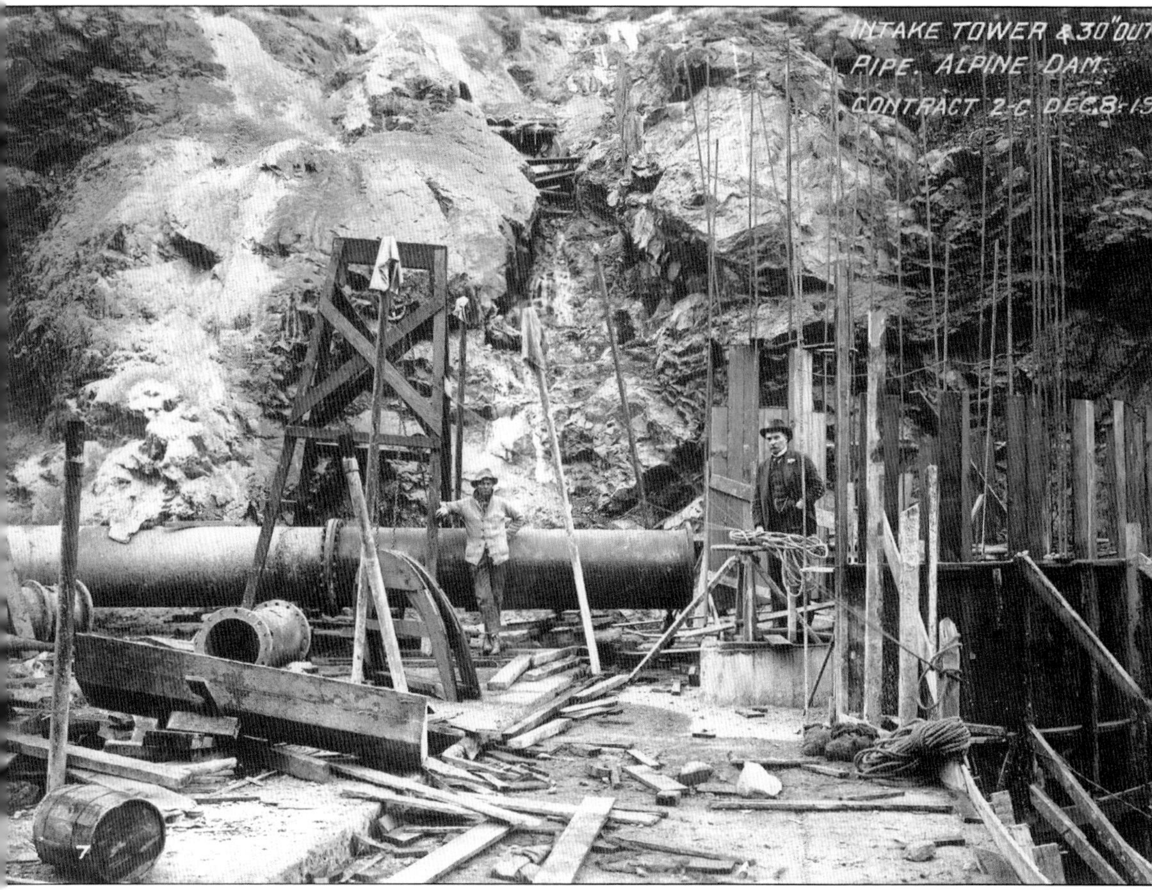

The starting point of the main supply line located near the base of Alpine Dam was just being assembled when this photograph was taken on December 8, 1917. Michael O'Shaughnessy (right), who was a district board member and head of the engineering committee, surveys the project. O'Shaughnessy had long been familiar with the plan for a dam at that site, as he had twice before made studies of a proposed dam there. Hired in 1906, and again in 1909 by F.P. Howard, president of the early Lagunitas Water Company, O'Shaughnessy made recommendations for that location as a reservoir site.

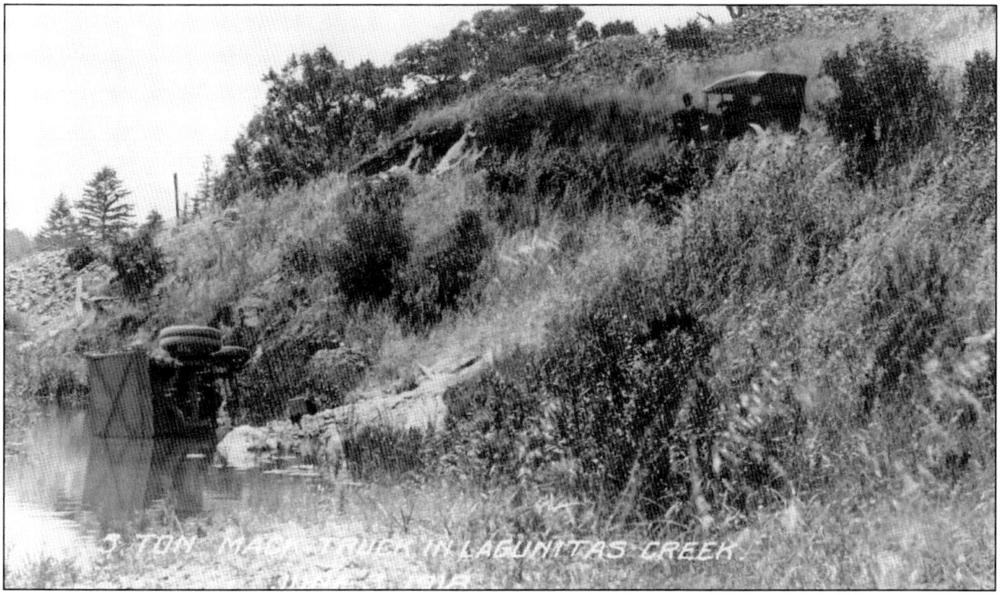

As MMWD took over completion of Alpine Dam from the contractor, equipment problems became apparent, as the contractor's equipment proved to be worn out and hard to keep running. This photograph shows a May 31, 1918, incident when a work truck ran off the road and into Lagunitas Creek, seriously injuring the driver and slightly injuring three other workers.

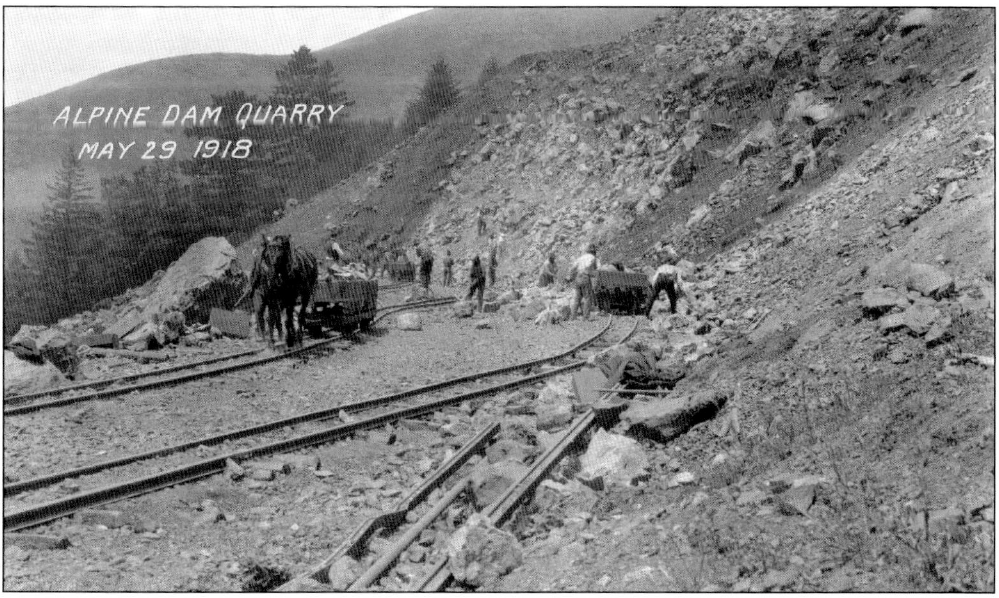

In order to avoid transportation costs, a quarry was opened near the north side of Alpine Dam. It provided large quantities of gravel that were mixed with cement to form the concrete for dam construction. In addition to dynamite, excavation was performed by pick and shovel with the horse-drawn rail carts shown here.

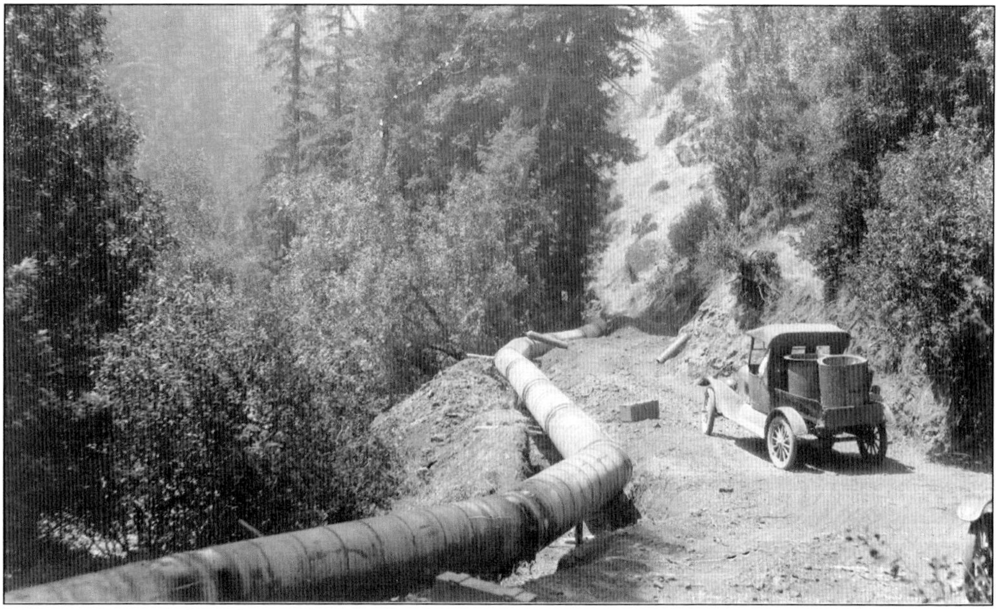

The concrete pipeline from Alpine Dam was built in two stages, with the first leading to the west portal of Pine Mountain Tunnel and the second from the east portal of the tunnel. When finished, the pipeline passed over 40 concrete trestles supporting the pipe, which was 24 inches in diameter and made of reinforced concrete. The rough terrain of Mount Tamalpais made construction of the concrete pipeline difficult. The sections of pipe were required to be short in order to keep the weight manageable when transporting them to the remote work sites. Construction yards, like the Fairfax yard seen below, were required for storing the large supply of pipe and other equipment.

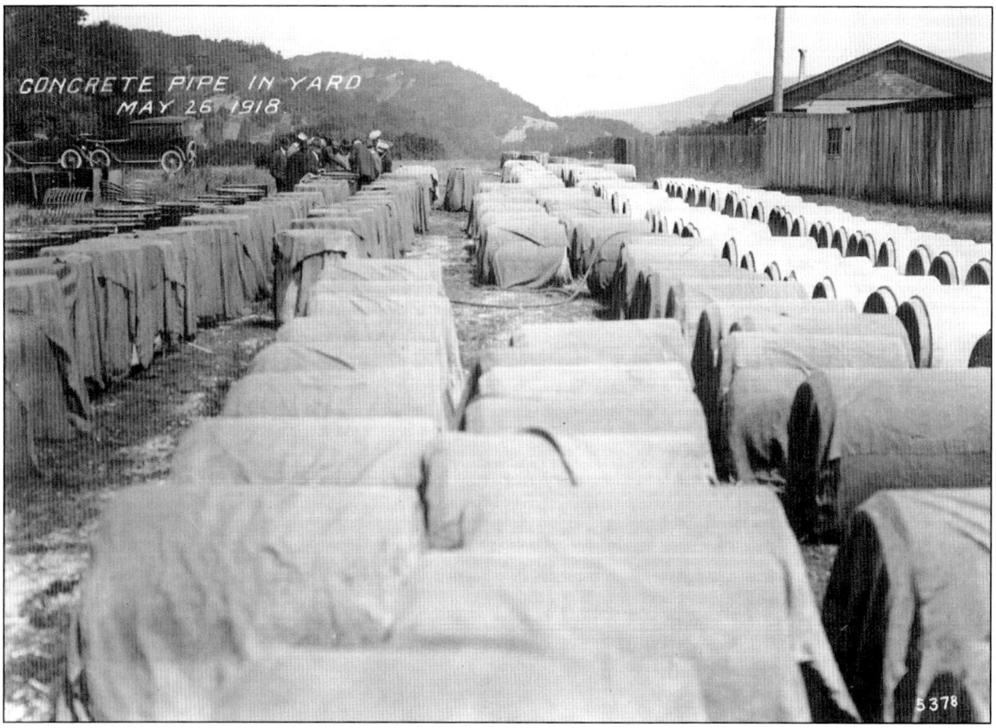

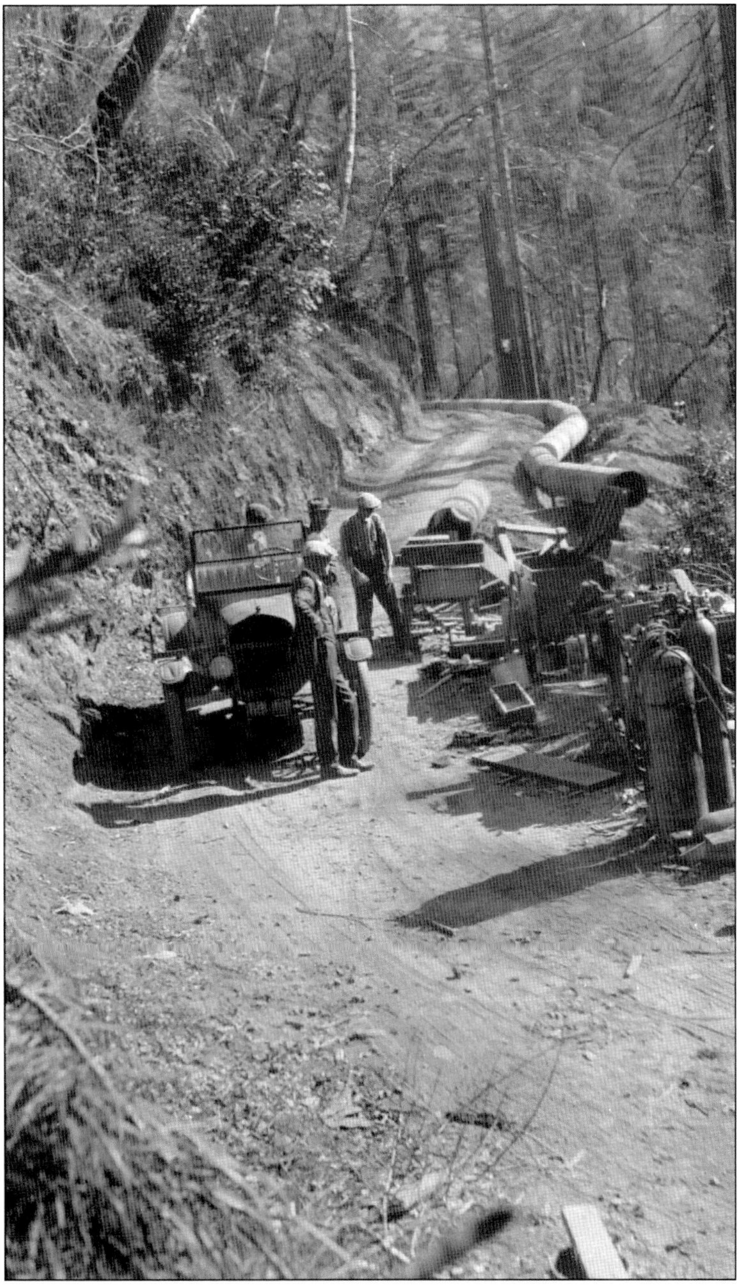

Alpine Dam received much of the publicity, but it was only one part of the Alpine Gravity System. The system was made up of the dam, the concrete pipeline, and Pine Mountain Tunnel. At the time of original construction, water left the dam through the concrete pipeline transmission system, which wound at a uniform grade of one foot in 1,000 for 3.5 miles, down the Lagunitas Canyon to Little Carson Creek, where it emptied into Pine Mountain Tunnel. Upon leaving the tunnel at the east portal, above the Cascade area of Fairfax, the water entered the second section of the concrete transmission pipeline. After supplying the Fairfax area, the concrete transmission pipeline continued eastward through Porteous Tunnel and on to the Phoenix Lake system. In 1919, the delivery capacity of the pipeline was five million gallons of water in a 24-hour period.

DEDICATION OF ALPINE DAM

Saturday, June 21, 1919

At 3 O'CLOCK P. M.

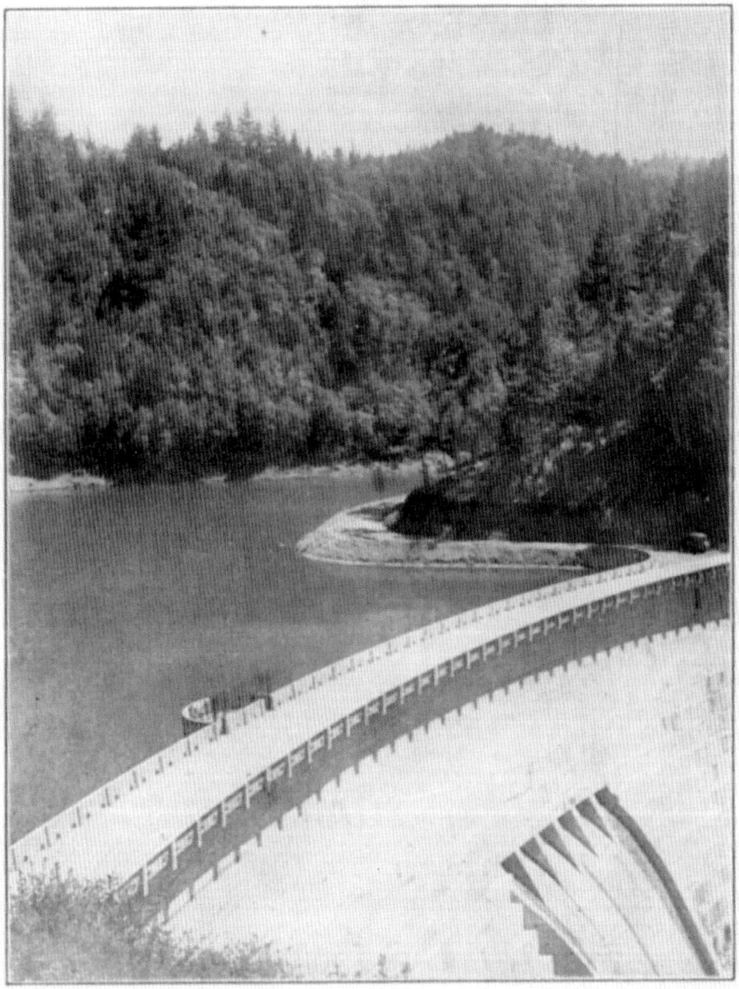

Built 1918, by the Marin Municipal Water District

A. R. BAKER,
Engineer.

M. M. O'SHAUGHNESSY,
Consulting Engineer.

On Saturday, June 21, 1919, the Marin Municipal Water District held a dedication ceremony for the newly constructed Alpine Dam. The materials handed out at the ceremony described the dam and what it would mean for the community. At the bottom, the dam itself was 78 feet thick, and the foundations extended 15 feet into the solid rock below the creek bed, and on 15 feet on the sides. The dam contained 27,719 cubic yards of concrete, which was poured from the mixing plant built on the side of the hill north of the dam. The water surface of Alpine Lake was 135 acres and at that time it held 1,017,300,000 gallons of water. Those present were told that "the system will furnish an unlimited supply of water for many years to come." Despite all the problems encountered by the original contractors, all the work and property acquisitions proposed in the bond issue ran over by only $55,000.

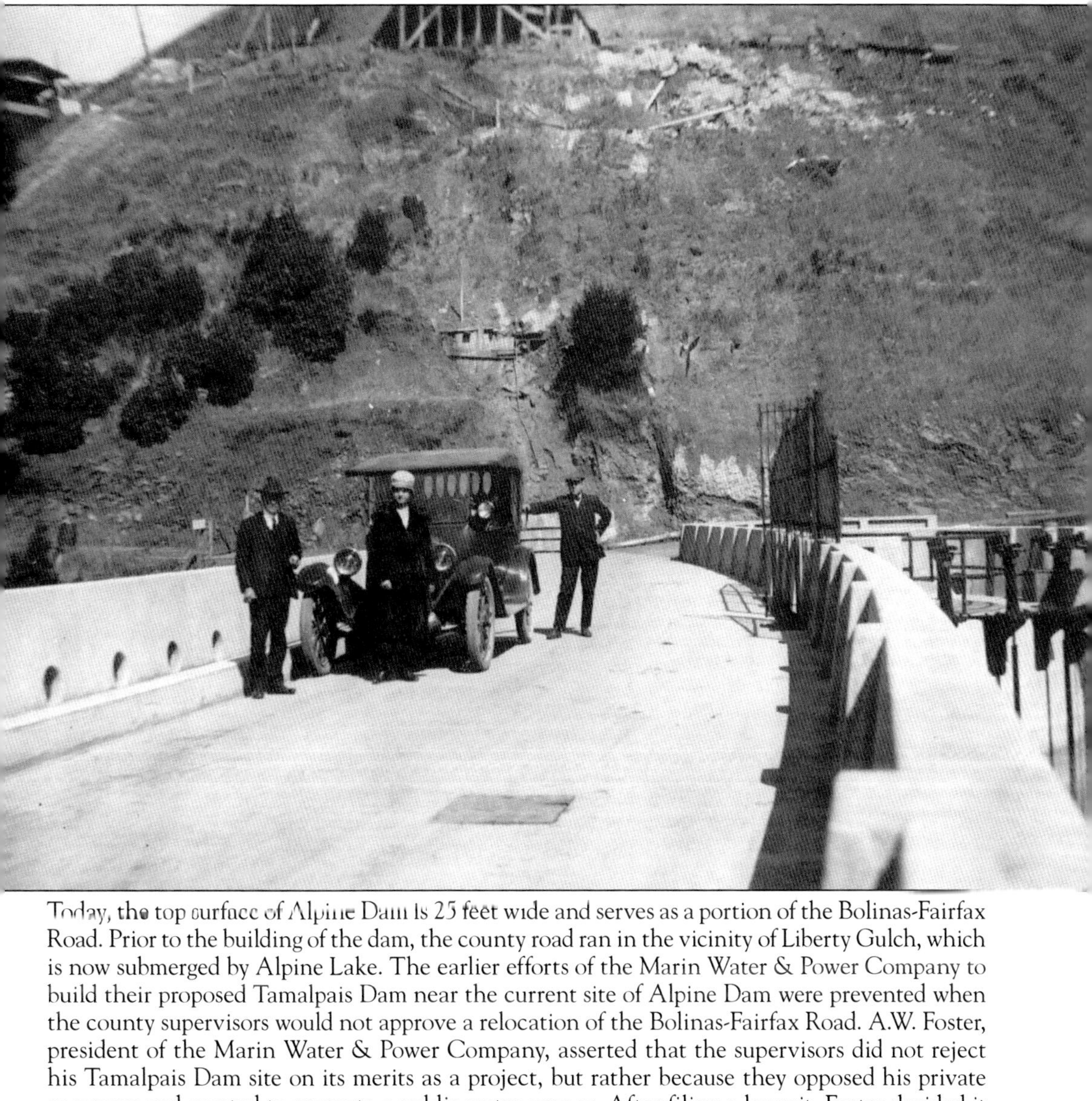

Today, the top surface of Alpine Dam is 25 feet wide and serves as a portion of the Bolinas-Fairfax Road. Prior to the building of the dam, the county road ran in the vicinity of Liberty Gulch, which is now submerged by Alpine Lake. The earlier efforts of the Marin Water & Power Company to build their proposed Tamalpais Dam near the current site of Alpine Dam were prevented when the county supervisors would not approve a relocation of the Bolinas-Fairfax Road. A.W. Foster, president of the Marin Water & Power Company, asserted that the supervisors did not reject his Tamalpais Dam site on its merits as a project, but rather because they opposed his private company and wanted to promote a public water agency. After filing a lawsuit, Foster decided it was better to abandon the location in favor of Phoenix Gulch, where he built Phoenix Dam. In 1913, as MMWD was planning for Alpine Dam, the district obtained permission from the county supervisors to close the Bolinas-Fairfax Road and relocate it across the top of the dam.

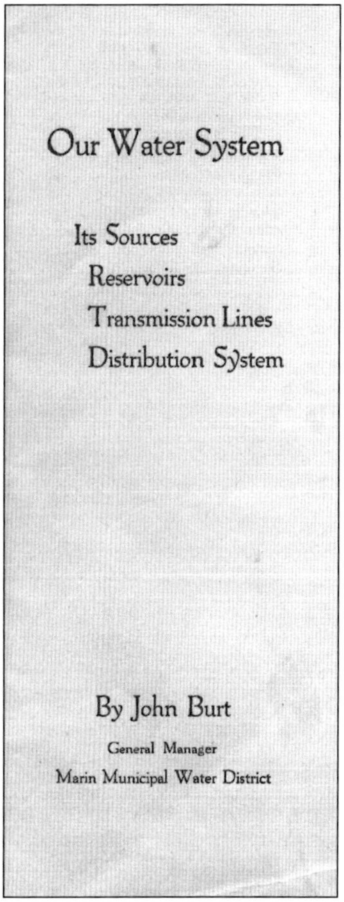

Our Water System

Its Sources
Reservoirs
Transmission Lines
Distribution System

By John Burt

General Manager
Marin Municipal Water District

By 1925, the assets of the private water companies had been incorporated into the MMWD water system. In April of that year, the water district requested approval of a $1.5-million bond issue, which the voters rejected. Concluding that the voters were not given enough information, the directors requested a new measure be brought before the voters in October 1925. The district proposed to expand the water supply with improvements, including a proposal to extend the height of Alpine Dam by 30 feet and thus expand the reservoir. The Publicity Committee produced more information, including the pamphlet pictured left and below. That pamphlet proved to be popular and was first read by general manager John Burt in a presentation before the San Rafael Rotary Club on September 28, 1925. In October, the bond measure passed with 86 percent of the vote.

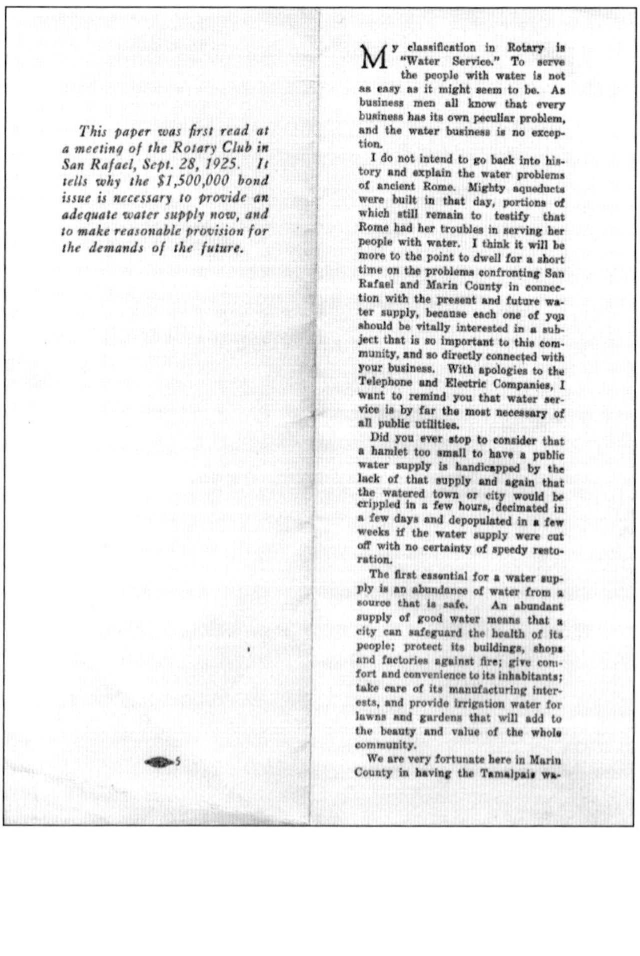

This paper was first read at a meeting of the Rotary Club in San Rafael, Sept. 28, 1925. It tells why the $1,500,000 bond issue is necessary to provide an adequate water supply now, and to make reasonable provision for the demands of the future.

5

My classification in Rotary is "Water Service." To serve the people with water is not as easy as it might seem to be. As business men all know that every business has its own peculiar problem, and the water business is no exception.

I do not intend to go back into history and explain the water problems of ancient Rome. Mighty aqueducts were built in that day, portions of which still remain to testify that Rome had her troubles in serving her people with water. I think it will be more to the point to dwell for a short time on the problems confronting San Rafael and Marin County in connection with the present and future water supply, because each one of you should be vitally interested in a subject that is so important to this community, and so directly connected with your business. With apologies to the Telephone and Electric Companies, I want to remind you that water service is by far the most necessary of all public utilities.

Did you ever stop to consider that a hamlet too small to have a public water supply is handicapped by the lack of that supply and again that the watered town or city would be crippled in a few hours, decimated in a few days and depopulated in a few weeks if the water supply were cut off with no certainty of speedy restoration.

The first essential for a water supply is an abundance of water from a source that is safe. An abundant supply of good water means that a city can safeguard the health of its people; protect its buildings, shops and factories against fire; give comfort and convenience to its inhabitants; take care of its manufacturing interests, and provide irrigation water for lawns and gardens that will add to the beauty and value of the whole community.

We are very fortunate here in Marin County in having the Tamalpais wa-

Four

The Mountain in Mid-Century

Mount Tamalpais was at the center of activity by the middle of the 20th century. In 1933, President Roosevelt began one of his most popular New Deal programs, the Civilian Conservation Corps (CCC). With millions of Americans unemployed, the CCC put young men and veterans to work on public works programs. In October 1933, a group of World War I veterans formed a temporary camp at Lake Lagunitas. A more permanent camp was built at Muir Woods, and by 1935, a second Marin camp staffed by young men, mostly from the eastern and southern states, was built across the road from Camp California—the University of California's College of Civil Engineering's Summer School of Surveying (near the Meadow Club fairways and MMWD's present-day Sky Oaks Watershed Headquarters in Fairfax). Over the six or seven years the CCC worked on the mountain, the men built the picnic grounds at Lakes Lagunitas and Phoenix; many spring-fed redwood fire tanks; miles of fire breaks and trails; bridges and other improvements to the streams and banks.

The CCC work was over by the 1940s, but the mountain soon took on a new importance. On Monday, December 8, 1941, just hours after the attack on Pearl Harbor, the MMWD Board of Directors met in special session. The security of the dams and the waterworks was the main topic that night. But soon that concern was coupled with the larger challenge of security for the entire West Coast. As early as July 1942, the captain of the US Army Corps of Engineers had contacted MMWD general manager James Peters and sought permission to use the west peak of Mount Tamalpais. By November, the Army was using the peak for airplane interception, using experimental radar equipment. Following the outbreak of the war, the top of the mountain was used by the Army exclusively and public access was cut off. Two local companies of the California State Guard were stationed at the old CCC Camp near Lake Lagunitas and put in charge of security of the dams and reservoirs. The military presence on the west peak remained for many years.

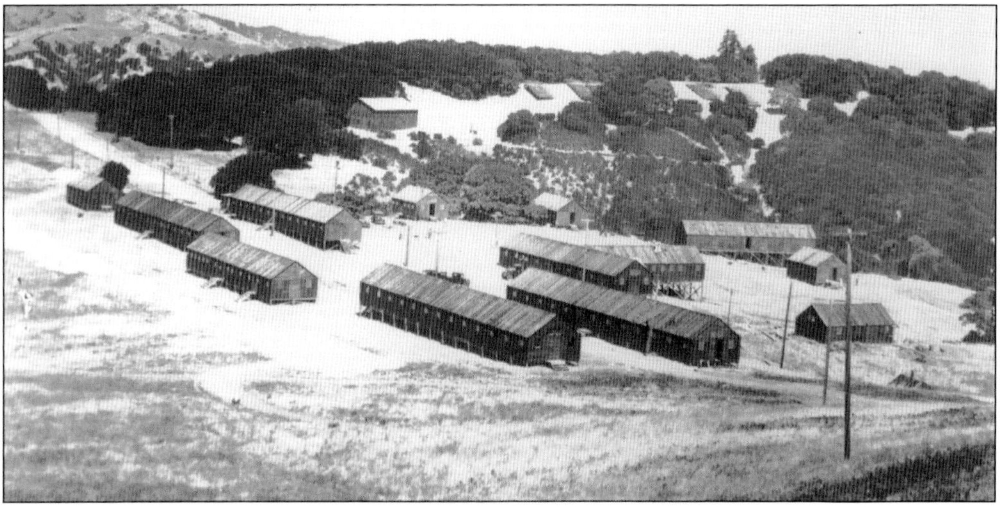

The CCC built many recreational facilities on the watershed, including Camp Alpine (above), built along the old county road, approximately 200 yards south of the present-day Sky Oaks Watershed Headquarters and ranger station. The camp, staffed by about 200 men, was completed in 1935 and operated year-round until it was closed in mid-1938. In that three-year period, the CCC also constructed the Mountain Theater, a fish hatchery at Lake Lagunitas, and picnic grounds at Lake Lagunitas (shown below), Deer Park, and Phoenix Lake. During this period, the corps also built 40 spring-fed redwood fire tanks, seven miles of fire trails, nine miles of truck roads, five miles of foot trails, and two vehicle bridges. The dams along Lagunitas Creek and 2,000 yards of stream bank armoring and facing were also checked.

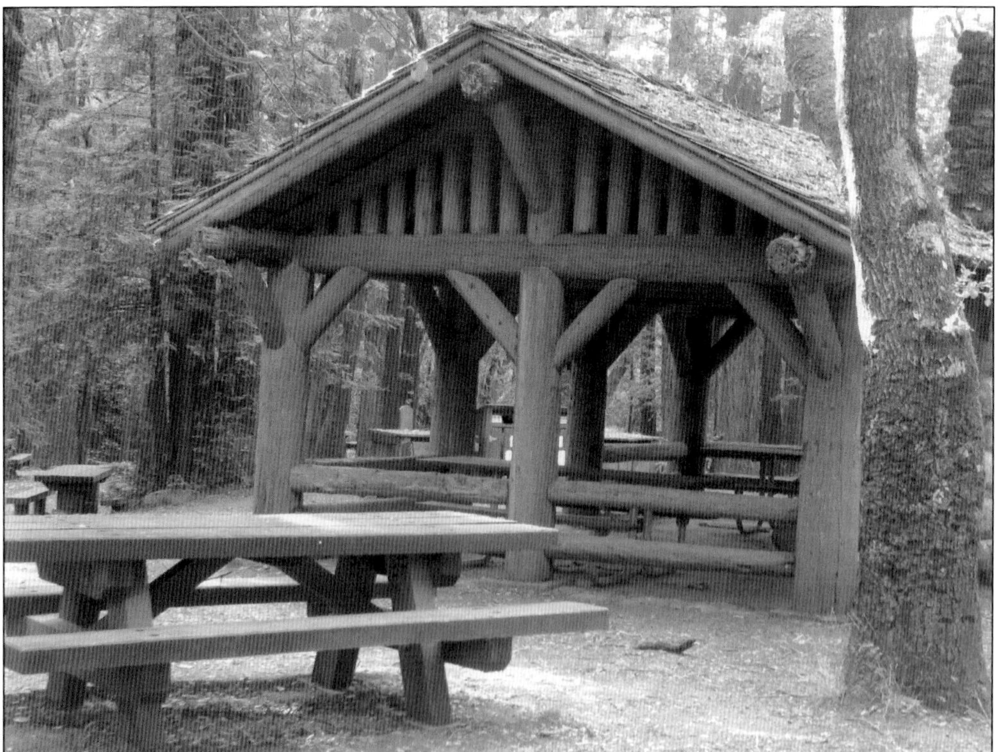

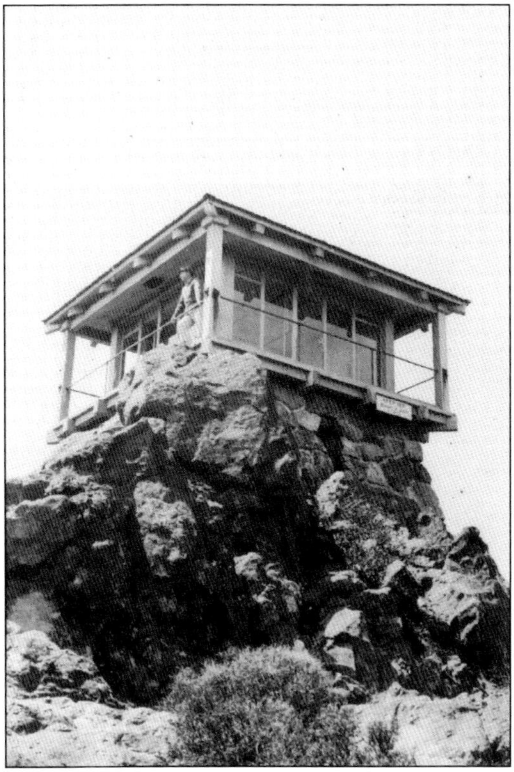

In 1935, the CCC built Gardner Lookout on the east peak of Mount Tamalpais. The lookout station was named after Edwin B. Gardner, the first fire chief of the Tamalpais Forest Fire District (TFFD). The fire district was created in 1917 by an act of the state legislature. The lookout station was built to replace an aging structure built by the *San Francisco Examiner* in 1901 to spot ships approaching San Francisco. The Marin County Fire Department, which replaced TFFD in 1941, continues to use the structure as a fire lookout today. Pictured below are Henry Long (left) and Fire Warden E.G. "Gus" Lanouette of the TFFD.

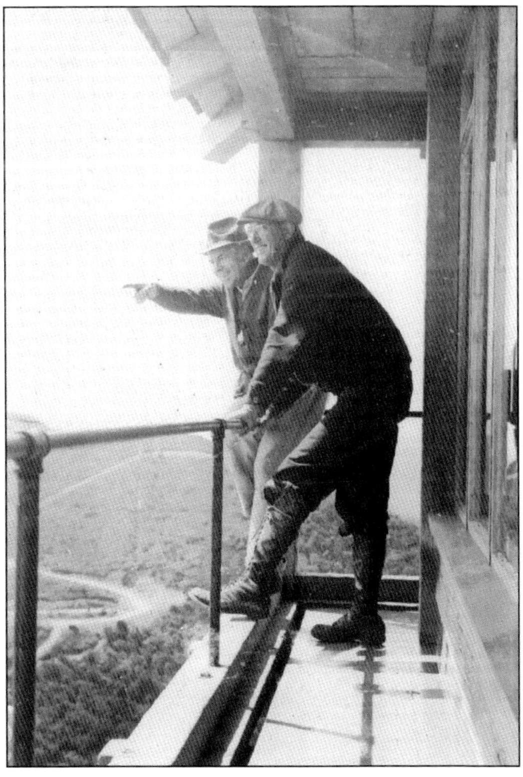

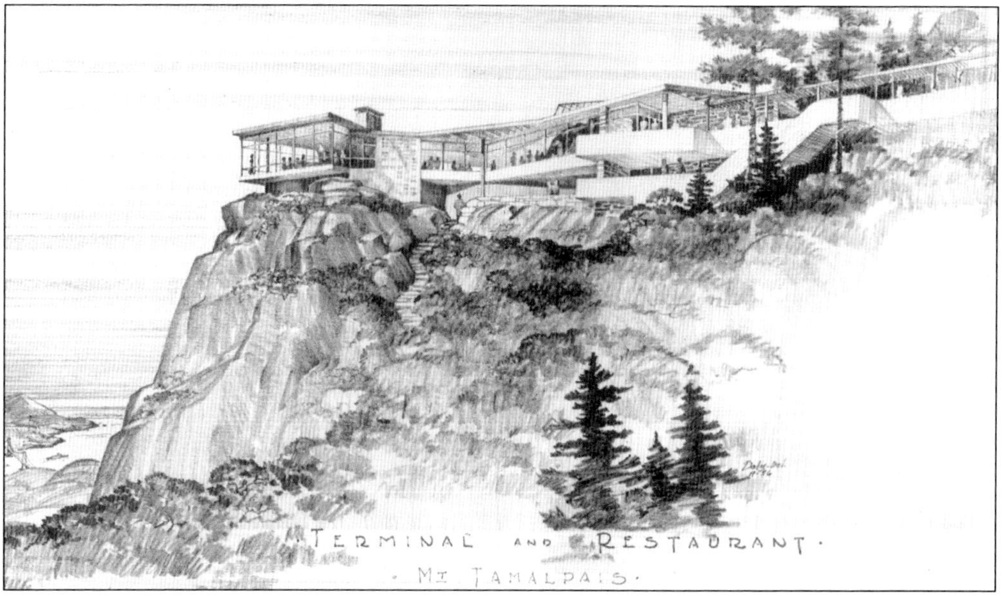

These sketches from the mid-1950s show a proposed development on the top of Mount Tamalpais that included an aerial tramway from Larkspur, a restaurant, and a gift shop. The tramway was modeled, in part, on a tramway on Cannon Mountain in Franconia, New Hampshire. The developer pointed out that the project would "bring added revenue, taxes, and national publicity to the community." The developer needed to get permission and access from MMWD to proceed with the project. Fortunately for the citizens of Marin, MMWD had the authority to deny permission for this and other proposed developments, keeping the mountain in a more natural condition for current and future residents.

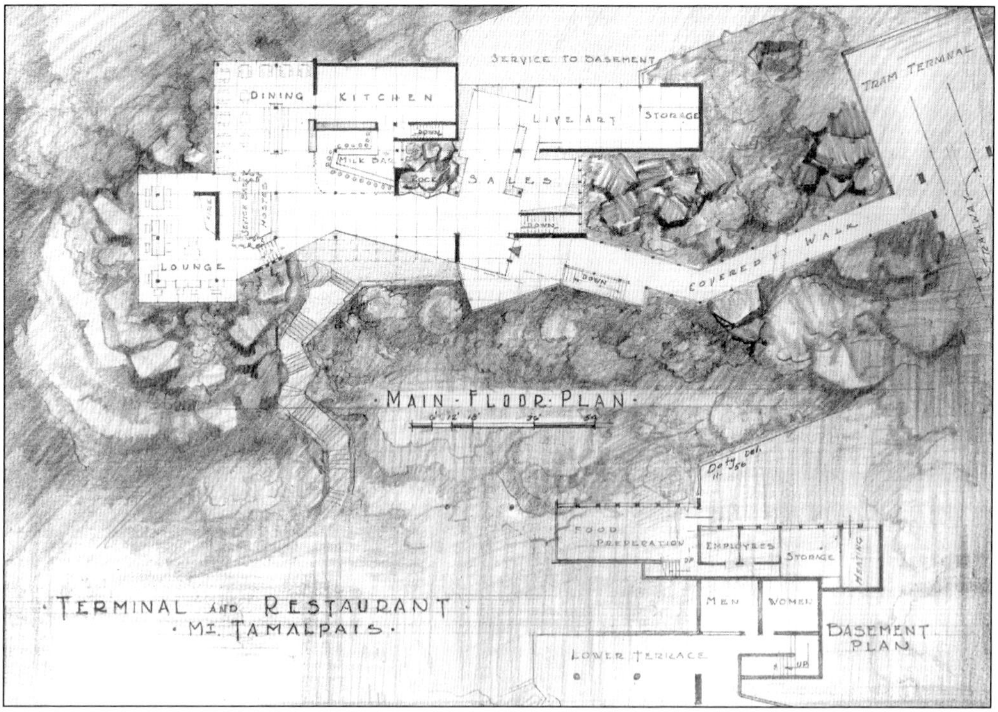

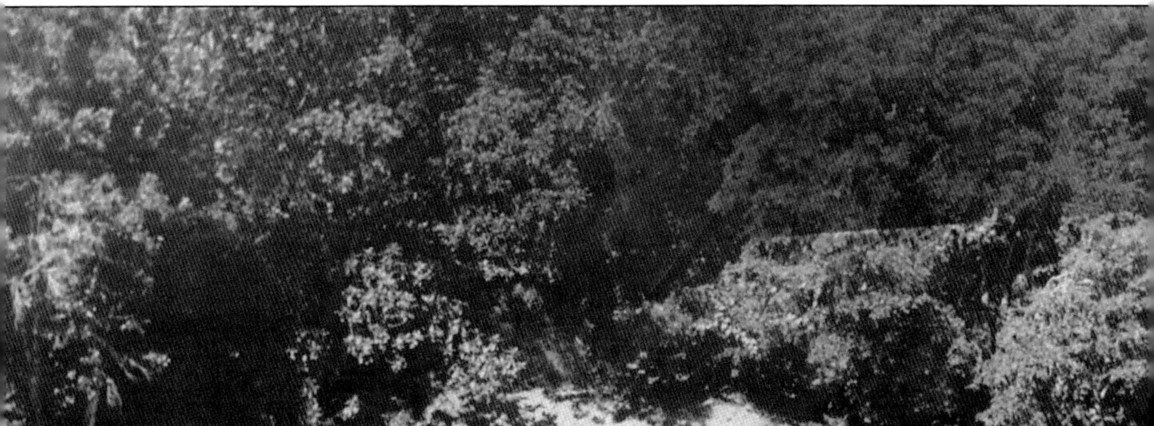

Fulfilling the promise of a park, which was made to Marin voters in 1912 as the water district was being created, the watershed and its reservoir areas have allowed public access and recreational activities. The rules and regulations governing activities on the watershed have changed over the years, recognizing MMWD's primary responsibility as a provider of clean water. Prior to World War II, the district allowed people to rent boats and row out onto the lakes. Today, boating is not allowed on the reservoirs, but this 1926 picture shows MMWD secretary Eben Prentice fishing on Lake Lagunitas. Prentice was appointed by the board of directors as secretary in December 1918, at a salary of $25 per month. He served in that position until his retirement in January 1935.

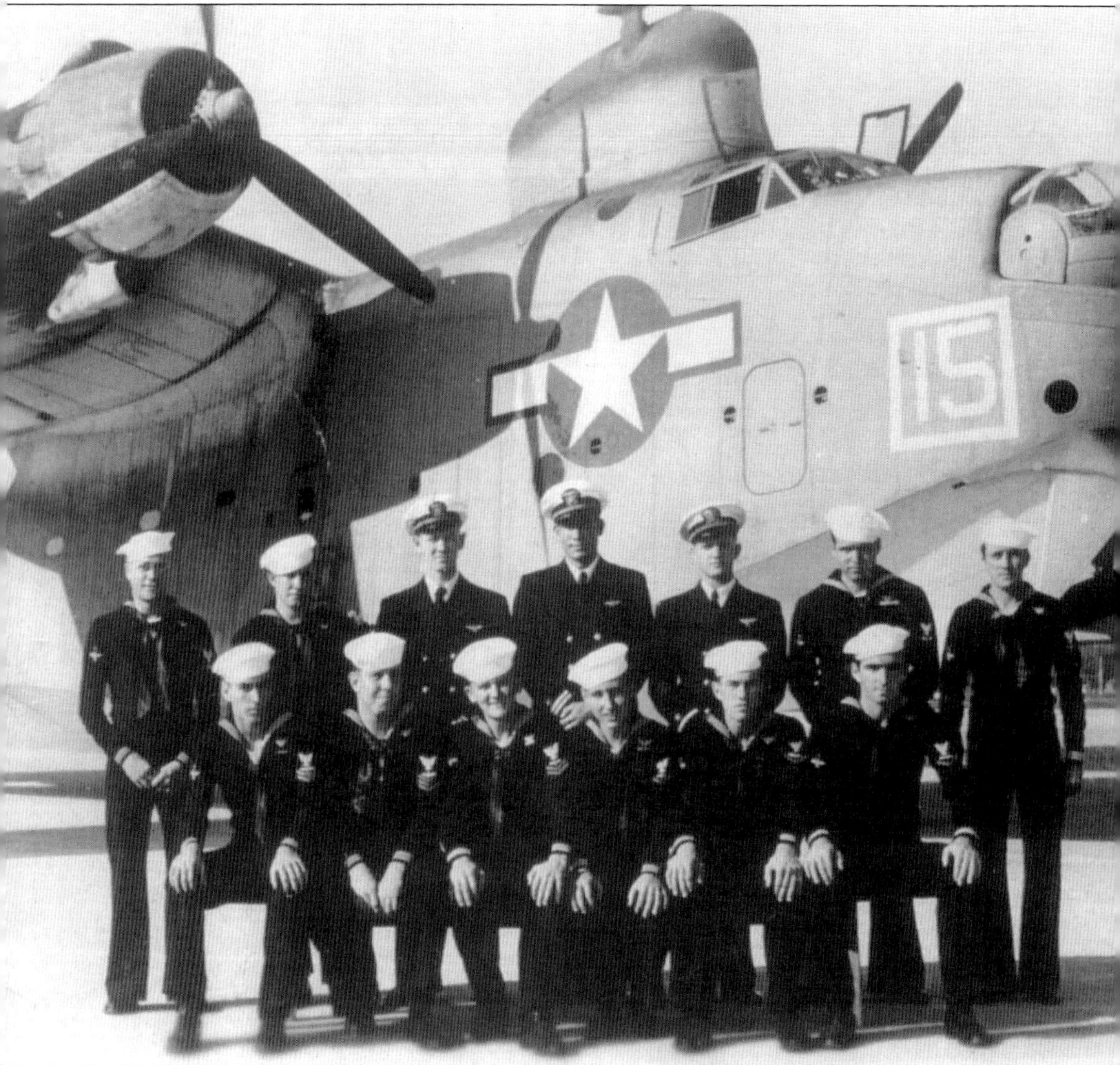

Combat Aircrew Charlie poses in front of Peter-15, a PBM-5 Mariner, a long-range patrol bomber, or "flying boat," used by the US Navy during World War II. Only two months after this photograph was taken, Combat Aircrew Charlie was assigned another PBM, Peter 9, and crashed it into the south side of Mount Tamalpais during a transpacific flight to Hawaii on November 30, 1944. They were attempting to return to Alameda Naval Air Station after suffering a right engine failure shortly after takeoff. Combat Aircrew Charlie had only eight members the night of the crash, all of whom were killed: Harry L. Holland, Aviation Chief Machinist Mate; Rodney Jeffers, Aviation Machinist Mate Second Class; Thomas J. Joyce Jr., Aviation Radioman Second Class; John R. Kelly, Aviation Radioman Third Class; Ensign Chapin B. Miller; Lieutenant (junior grade) Thomas W. Oliver Jr.; Wayne D. Paxson, Aviation Ordnance Man Second Class; and Lieutenant Joseph L. Resley. (Courtesy of Roger Bernier.)

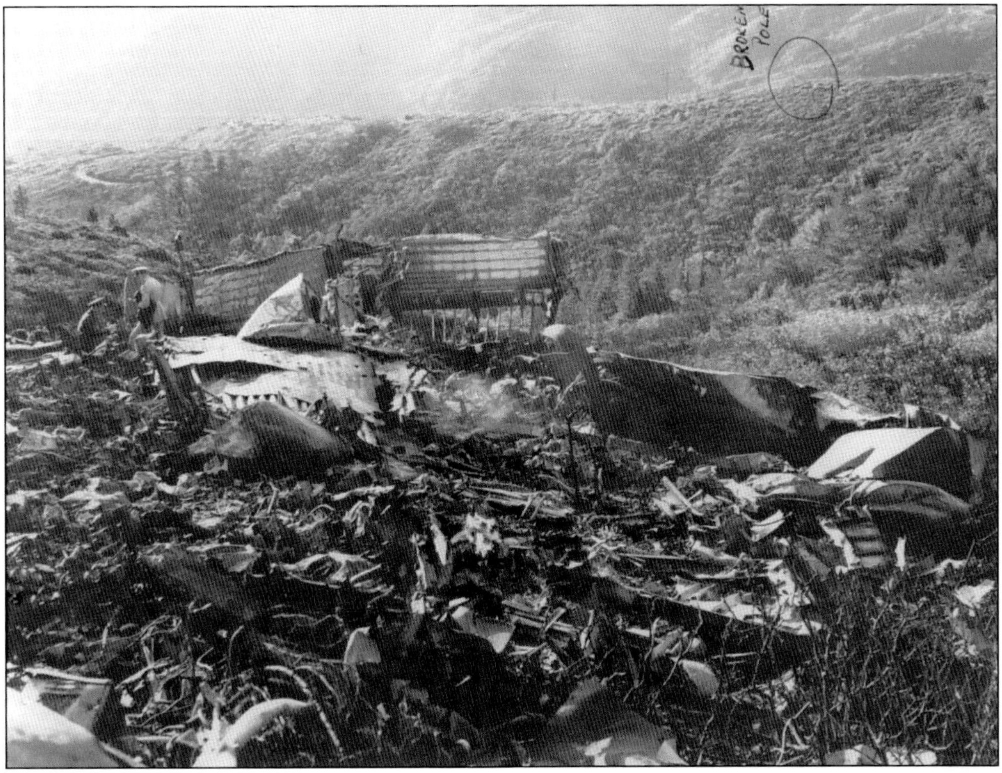

The Combat Aircrew Charlie crash site debris was scattered over a 300-yard area near the Double Bow Knot on the south side of Mount Tamalpais. Navy officers inspect the still-smoking wreckage. The largest pieces visible are remains of the tail section and wing. MMWD patrolmen, or "Water Rangers," as they were called in one newspaper article, were part of the recovery team. (Courtesy of Craig Fuller.)

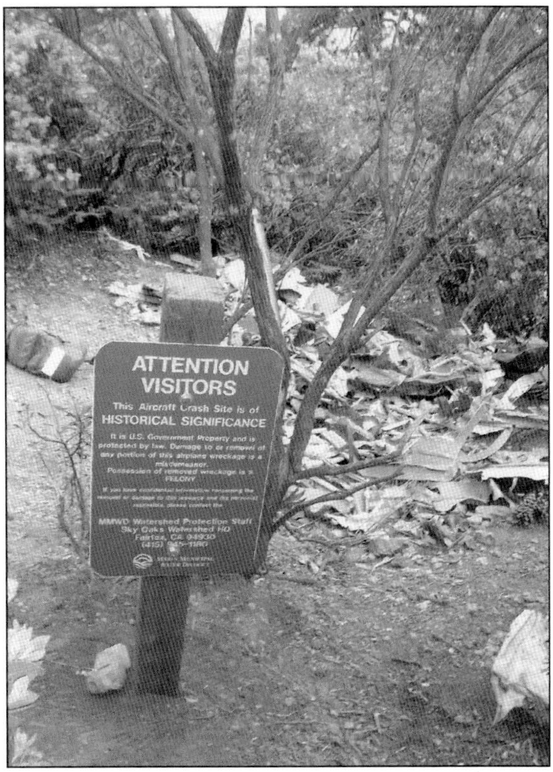

Some of the wreckage from the flight is still on the mountain. This sign posted by MMWD notifies visitors and warns against removal of debris. Shortly after World War II, on October 4, 1945, two Navy Corsair fighter planes collided over the Rock Spring area of Mount Tamalpais; both pilots survived although one was badly injured. An engine from one of the planes is still visible in Cataract Creek above Laurel Dell.

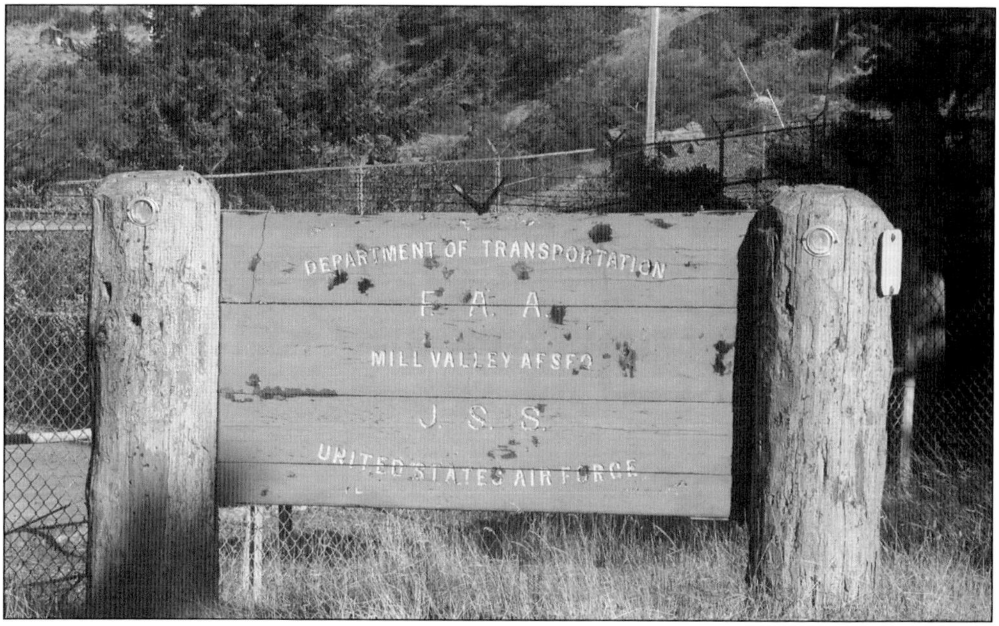

In 1941, MMWD was asked to allow the Air Corps Warning Service to use certain lands on the west peak of Mount Tamalpais for a detector site. After several meetings, the plan was formalized and the district agreed to allow 20 men with experimental radar equipment on the west peak for three weeks through December. With the outbreak of war on December 7, the Air Corps took on greater importance. Mount Tamalpais continued to be used for Air Force and other military needs until 2005, when the last federal government lease expired. Over the years, it was used for military communications as well as a permanent radar station, as part of the SAGE (Semi Automatic Ground Environment) network under the Air Defense Command. Although it no longer looks like it did in the 1980 photograph below, military remnants are still seen. (Below photograph by Len Rodriguez and pilot Sandy Mills.)

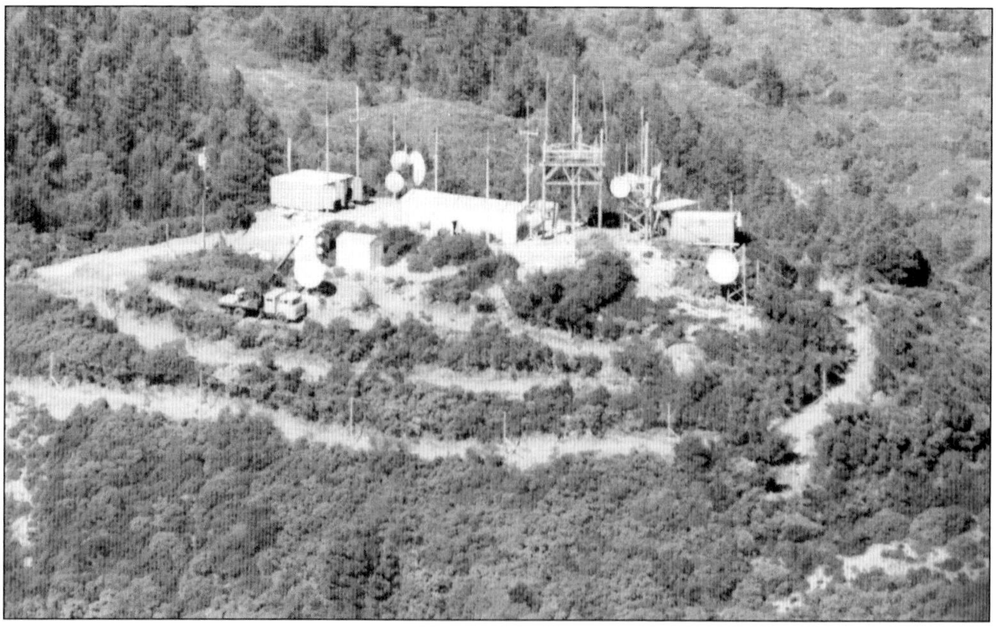

Five

Expanding the Water System

Shortly after completion of Alpine Dam, the California and Hawaiian Sugar Company (C&H) made an offer to purchase water from MMWD, hoping to deliver water to Point San Quentin and then send it by barge from the San Quentin pier to the C&H plant on the other side of the bay in Crockett. With a five-year purchase agreement, MMWD was able to build an expanded water transmission pipeline through the Ross Valley, fully paid for by C&H. The expansion was called the "Sugar Line." In 1924, in return for an extension of the contract, the sugar company paid for the raising of Alpine Dam by eight feet.

Between 1940 and 1942, the dam was raised again, this time by 30 feet. The thickness of the dam was increased on its downstream face by adding a 40-foot-thick slab of concrete to the base of the dam, which tapered in thickness as it rose to the top of the existing dam. The result was an expanded water system and a water supply predicted to be sufficient for Marin for many years.

However, from the end of World War II through the 1960s, Marin experienced a burst of growth. During this period, MMWD developed three more dams and reservoirs: Bon Tempe Dam and Reservoir in 1948, Peters Dam and Kent Lake in 1953, and Seeger Dam and Nicasio Reservoir in 1956. From 1958 to 1962, to keep up with changes in water quality standards, the district built both the Bon Tempe and San Geronimo treatment plants. In order to finance these projects, the voters approved bond measures in 1949 and 1956 totaling $16.6 million. Treatment improvements have continued throughout the decades since both plants were built. Today, these projects, along with Alpine Lake, constitute the heart of the MMWD water system, which now totals seven dams. By the end of the 1960s, the district was getting all of its water from Lagunitas Creek through the chain of connected lakes used as a gravity flow system. Since the 1960s, with the exception of drought years, the dams and lakes at Lagunitas and Phoenix have played little part in delivering water to consumers.

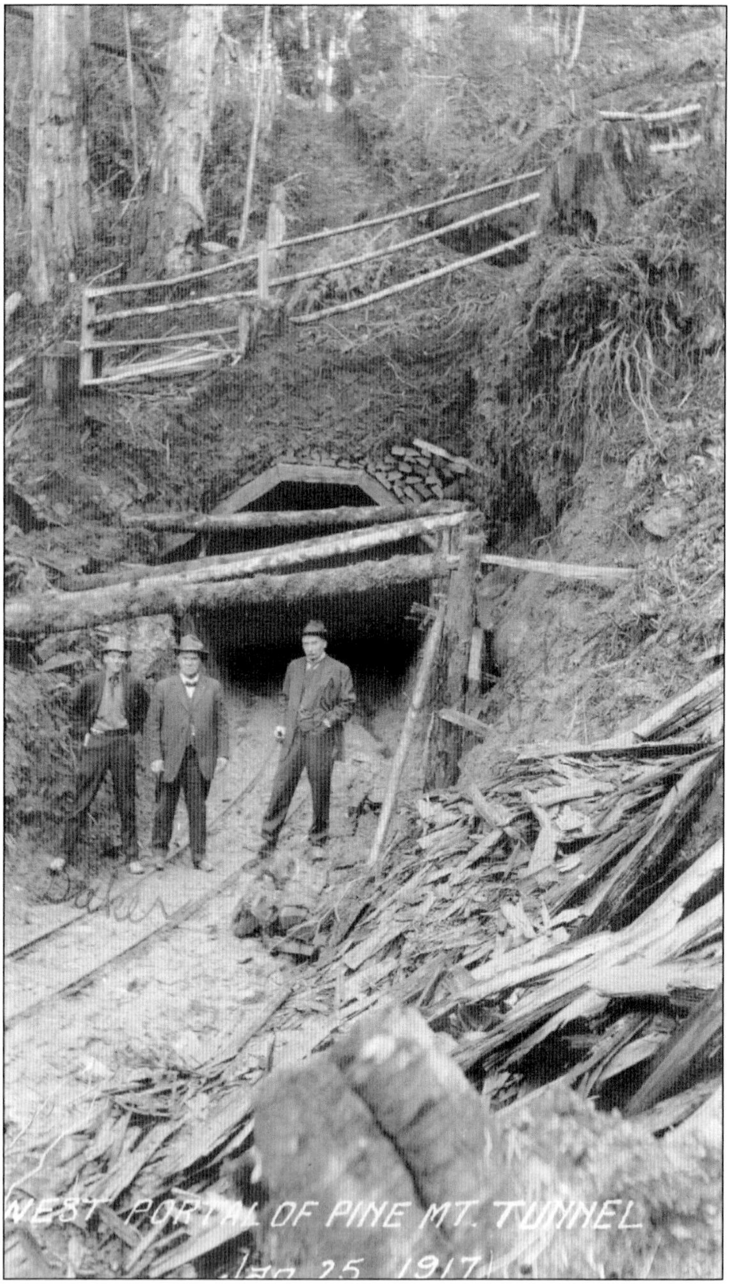

Pine Mountain Tunnel was built as part of the Alpine Dam Gravity System to convey water from Alpine Lake to Fairfax. This photograph was taken at the west end of the tunnel, located below Alpine Dam. MMWD engineer Albert R. Baker (left) and engineer Michael O'Shaughnessy (right) pose with an unidentified man. The tunnel is 900 feet below the summit of Pine Mountain ridge. The idea of building a tunnel through Pine Mountain did not originate with MMWD. As early as January 1868, the Mount Tamalpais Water Company proposed such a tunnel for an even larger project to supply the East Bay and San Francisco with water. A similar plan was proposed by the Lagunitas Water Company in 1909 to supply Richmond and the California and Hawaiian Sugar Company in Crockett.

The contract to build Pine Mountain Tunnel was awarded to McLaren and Peterson on their low bid of $257,400. Work started on December 5, 1916, and the tunnel was completed in 1918, after many problems. One 30-foot-long cave-in about 100 feet from the face of the east portal caused a serious delay. Other problems included encountering large bodies of natural gas, which in one instance caused an explosion, badly burning one man. Throughout 1917, there were labor problems at the site that slowed—and sometimes stopped—the project. At the June 7 MMWD board meeting, it was reported that work was being seriously undermined in the town of Fairfax by the unrestricted sale of bottled liquor to the workmen. A board resolution was passed asking the supervisors, in the name of good citizenship, to intervene.

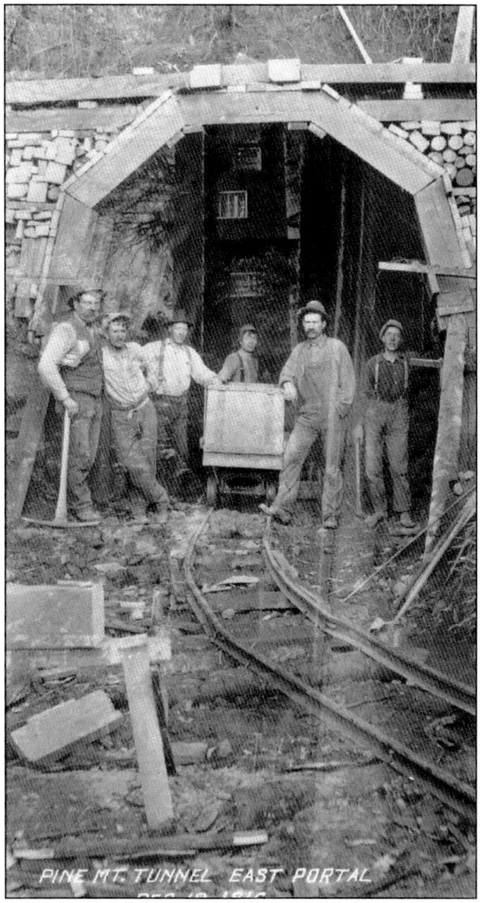

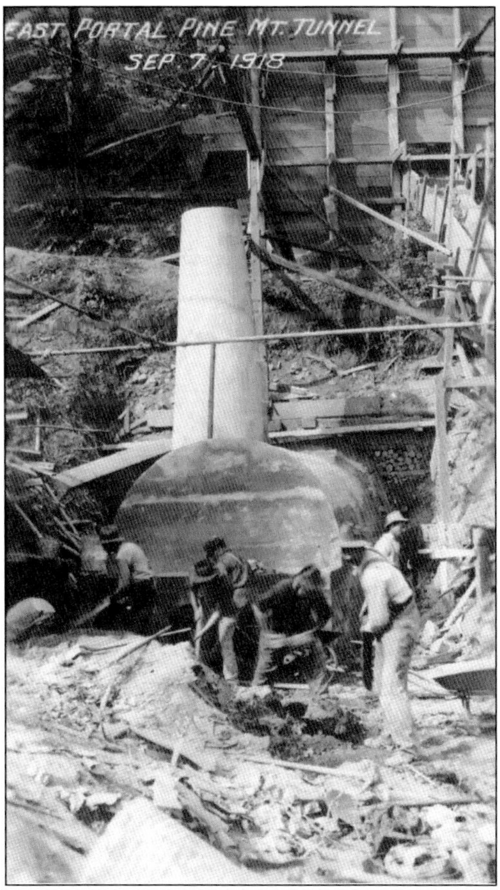

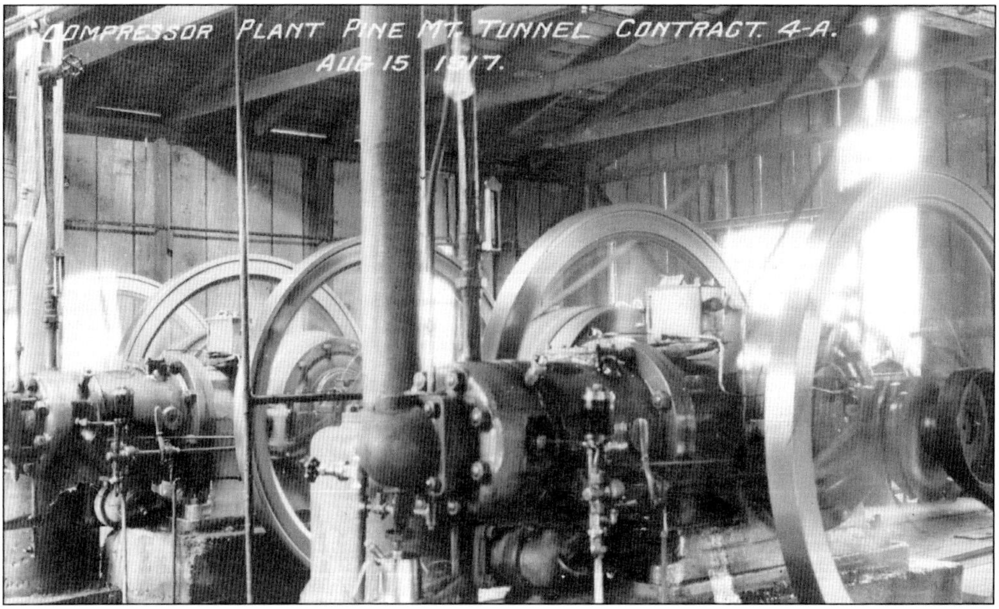

Three 120-cubic-foot compressors powered by two 25-horsepower gas engines, housed over the east portal of the tunnel, provided the power for the pneumatic drills to excavate Pine Mountain Tunnel. Lining up the two ends of the tunnels was an engineering and construction achievement; the alignment and grades checked almost perfectly, with 2/100ths variation in alignment and 15/1,000ths in grade.

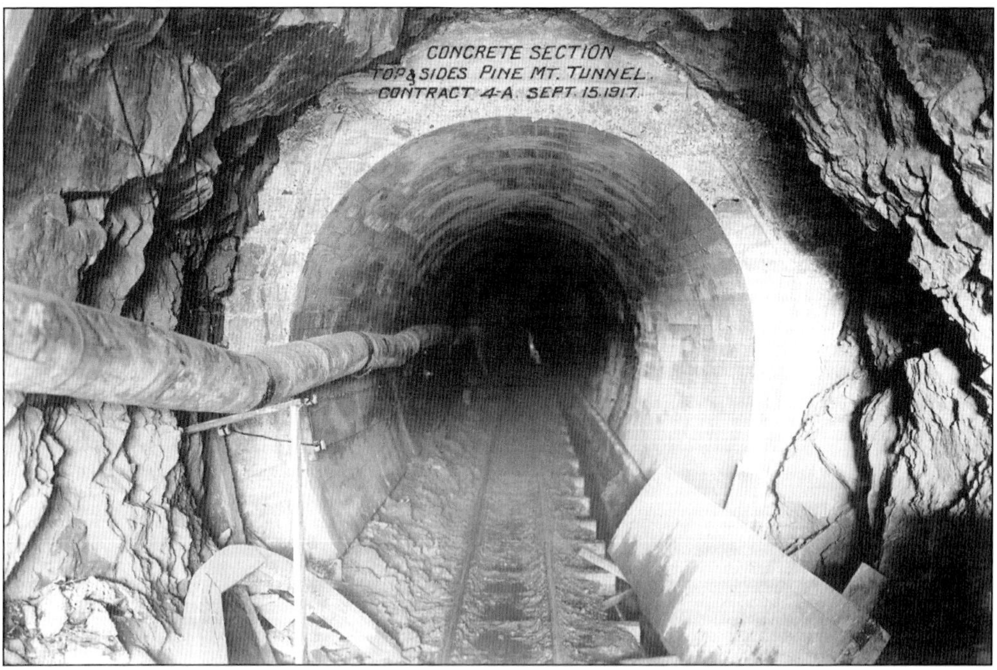

The thickness of the concrete lining of Pine Mountain Tunnel varies from six to 12 inches, depending on the hardness of the rock encountered. The tunnel is eight feet high, eight feet wide, and more than 1.5 miles, or 8,700 feet, long, with a capacity of three million gallons. It was completed in approximately two years at a cost of $287,515.90.

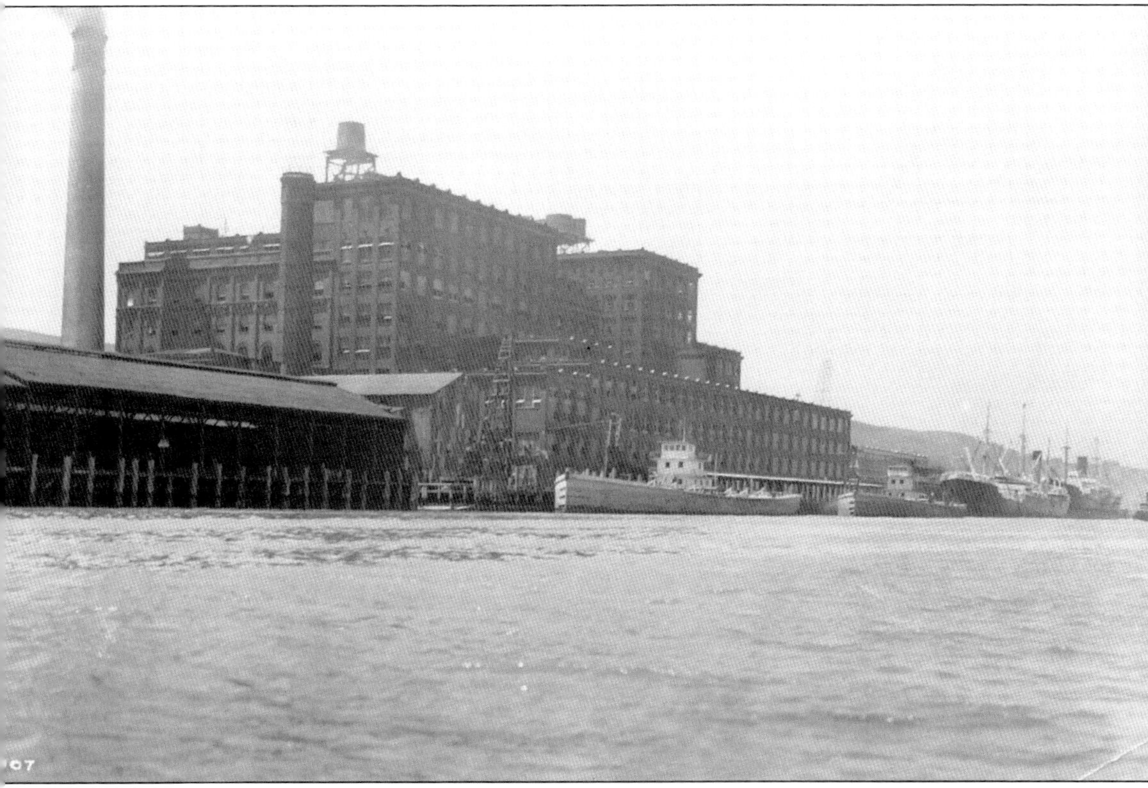

In 1924, MMWD signed a contract with C&H to sell fresh water to the company, which it needed to process sugar. Under the contract, MMWD agreed to supply water for C&H's processing plant in Crockett (above). In order to do that, the district agreed to raise Alpine Dam by eight feet and to lay six miles of 12-inch pipe from Phoenix Lake in Ross to Point San Quentin, on what became known as the "Sugar Line." C&H agreed to reimburse the district monthly for all expenses until the full cost was paid. A minimum yearly payment of $40,000 was guaranteed to the district under the contract. This agreement ran for approximately five years. The Sugar Line is still in use today as a water main supplying the Ross Valley.

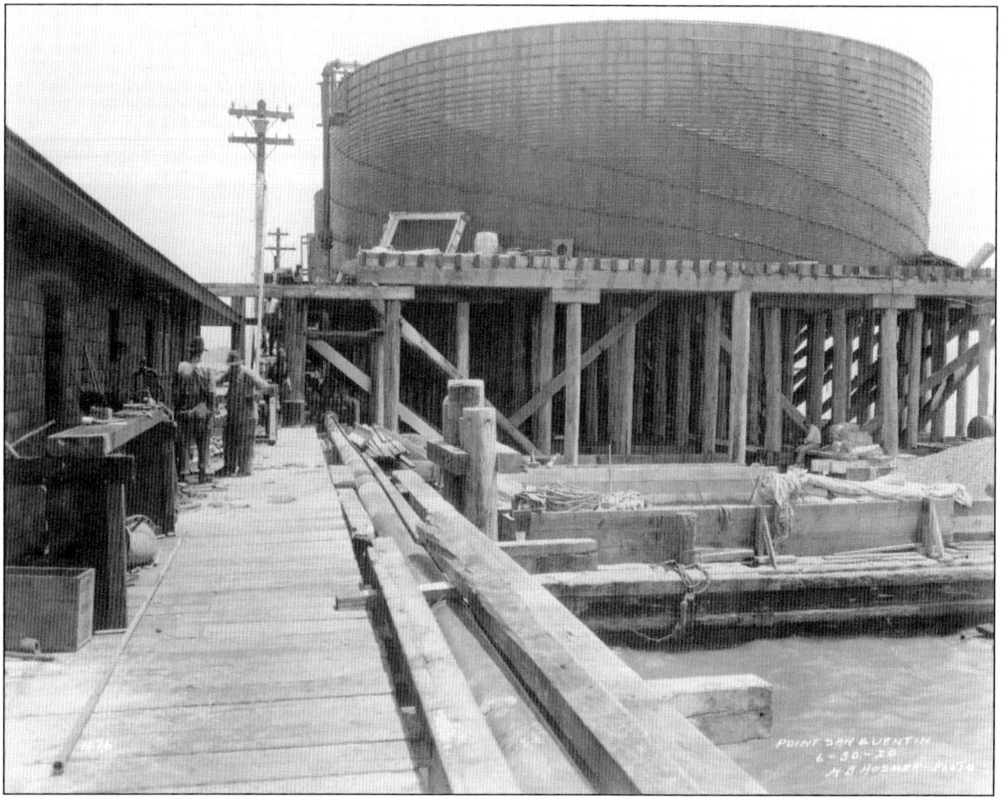

Under the terms of the contract, C&H agreed to pay MMWD $30,000 annually for delivery of 300 million gallons of water per year. It paid an additional $10,000 annually for another 200 million gallons, payable whether or not it took the extra water. Three 500,000-gallon redwood tanks were constructed by C&H near the end of its pier, for storage of the water delivered to them by barge. The above photograph, looking west in 1920, shows one of the tanks nearing completion. The below photograph shows the three completed tanks, with McNears Brickyard and quarry in the background. (Photographs by H.B. Hosmer.)

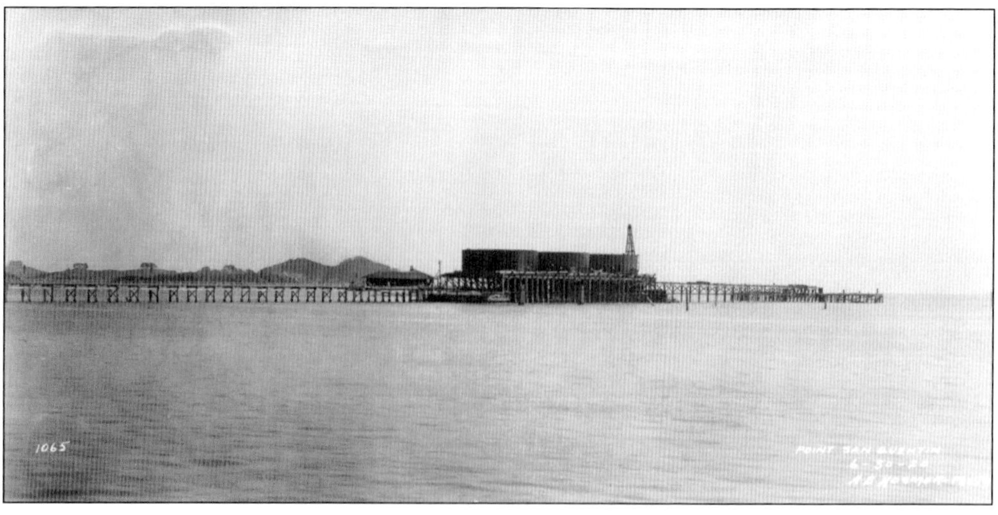

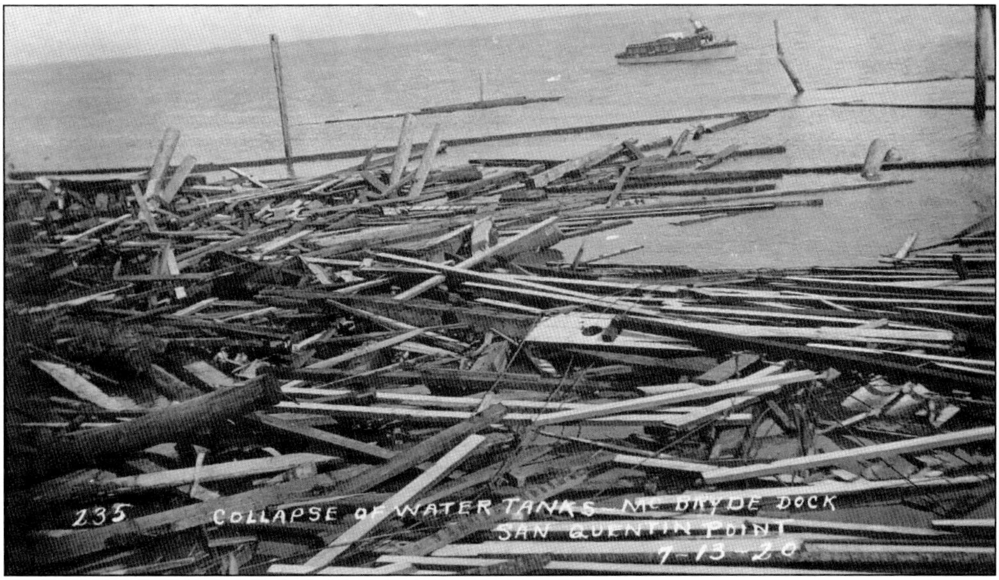

C&H completed building the redwood tanks on July 13, 1920. But when the tanks were filled with water for the first time, the weight was too much for the pilings, which may not have been sunk deep enough. All three tanks and part of the pier collapsed, leaving debris scattered over the entire area.

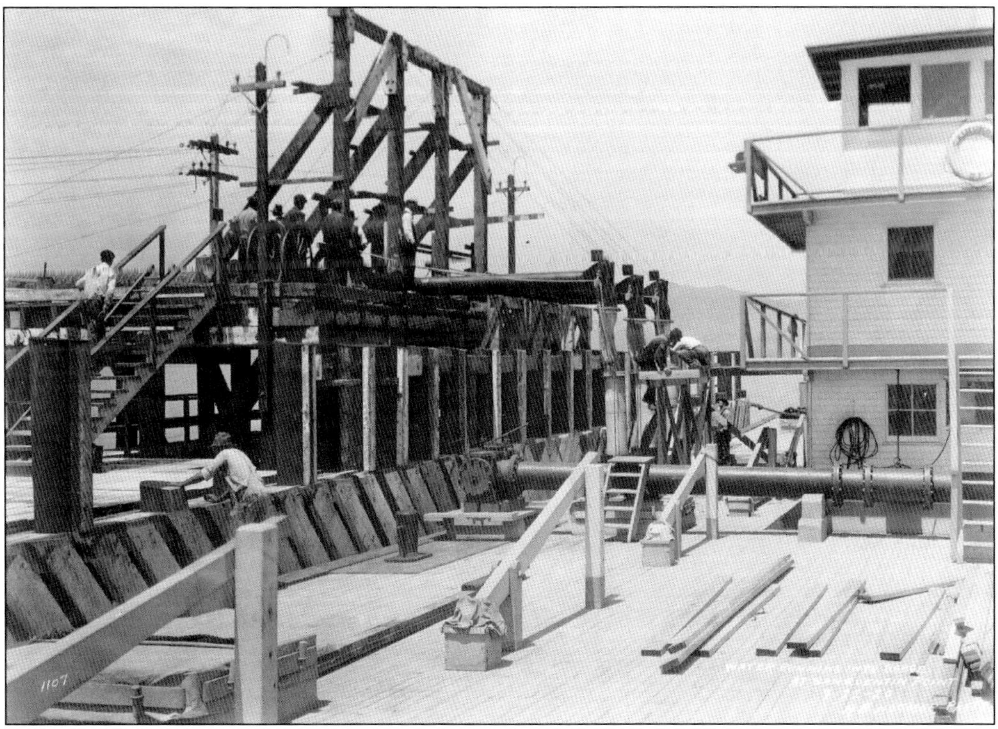

Within just a week, the debris was removed by C&H, and they resorted to "Plan B." Instead of rebuilding the three tanks, three large water spouts, similar to those used by railroads to fill steam engines, were positioned at the end of the pier to fill the barges with water directly from the MMWD system.

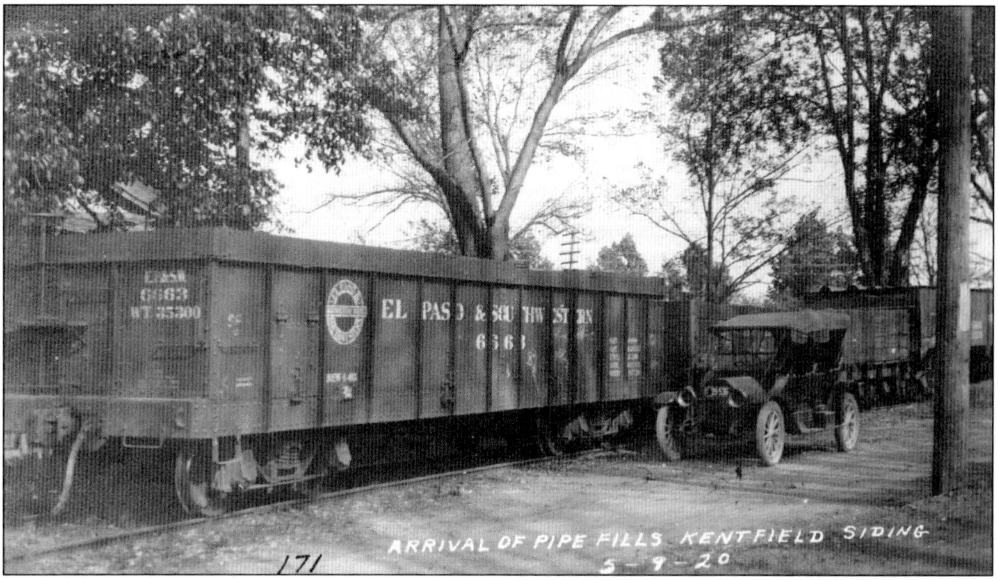

For construction of the Sugar Line, the water district ordered 33,250 feet of 12-inch cast-iron bell and spigot pipe from American Cast Iron Pipe Company in Birmingham, Alabama, with delivery to be made by railroad. Railroads played an important role in shipping pipe to different locations in Marin. Eight carloads of pipe were delivered to the Kentfield Station (above) and fourteen carloads to the Ross Station (below). Loads of pipe were also logistically placed at the Greenbrae and San Rafael Stations. Ten different railroad companies became involved in hauling the pipe to different spots in the county. The last delivery was made on June 13, 1920.

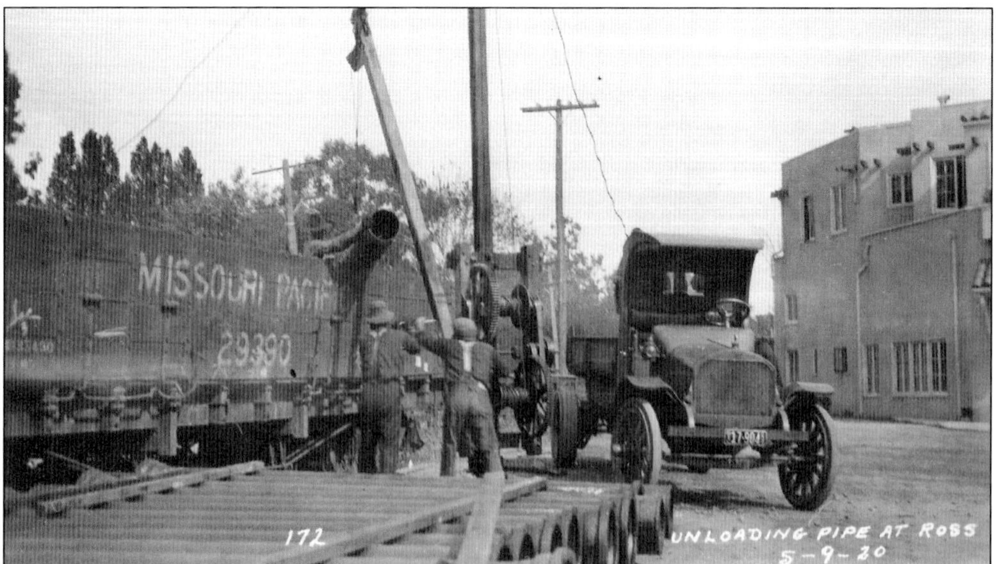

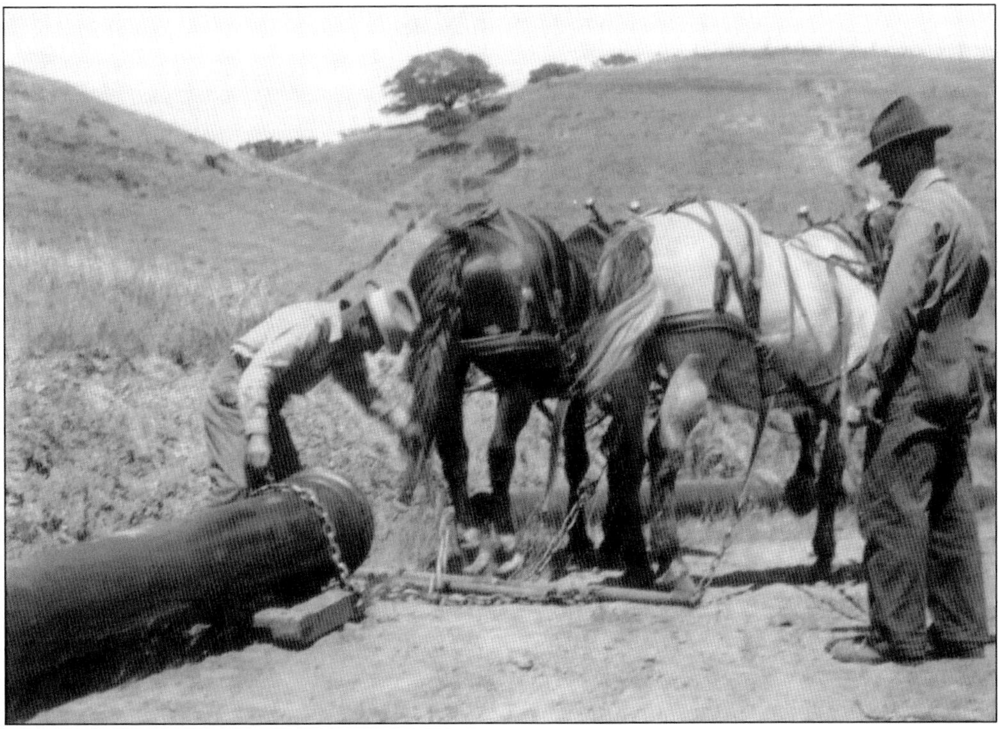

In 1920, laying long lines of pipe presented challenges. Each section of 12-inch cast-iron pipe was an enormous weight. Horses were the most efficient method for transporting the heavy cast-iron pipe off-road. This photograph, taken near San Quentin, shows the placement of pipes after they were hauled by a team of horses.

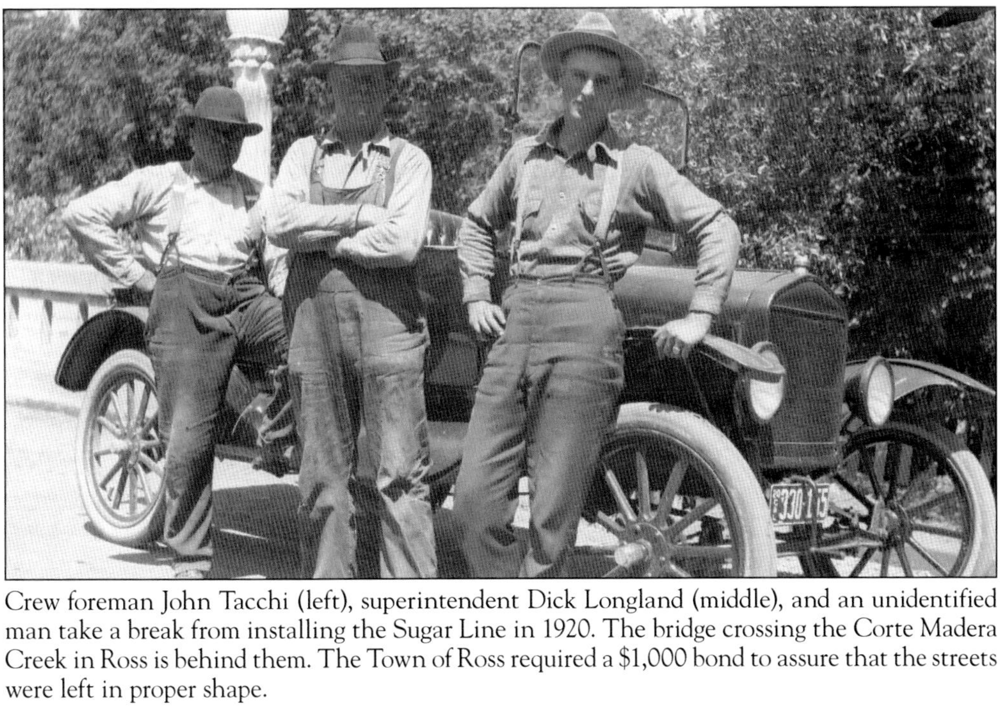

Crew foreman John Tacchi (left), superintendent Dick Longland (middle), and an unidentified man take a break from installing the Sugar Line in 1920. The bridge crossing the Corte Madera Creek in Ross is behind them. The Town of Ross required a $1,000 bond to assure that the streets were left in proper shape.

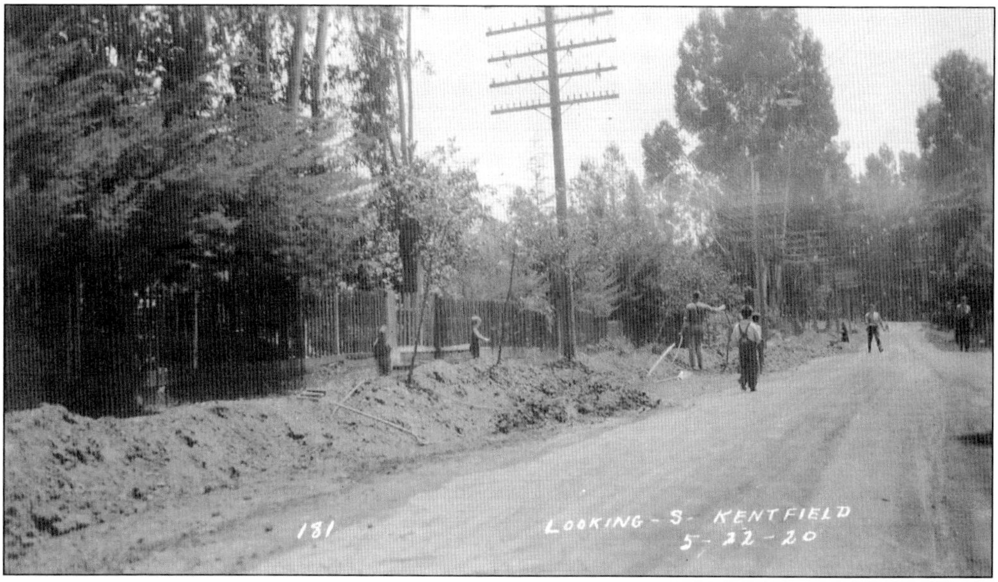

The Sugar Line began in Ross at the entrance of the MMWD property on Lagunitas Road. The line was set through Ross, Kentfield, and Greenbrae in May 1920. By June, work was being completed in San Rafael towards San Quentin. In the 1920s, pipeline trenches were dug by human labor, and all of the Sugar Line trenches had to be dug manually. The line was underway in Kentfield (above) and on Sir Francis Drake Boulevard (below). Today, the Sugar Line, built by MMWD at the expense of C&H, remains the main water line through much of Ross Valley.

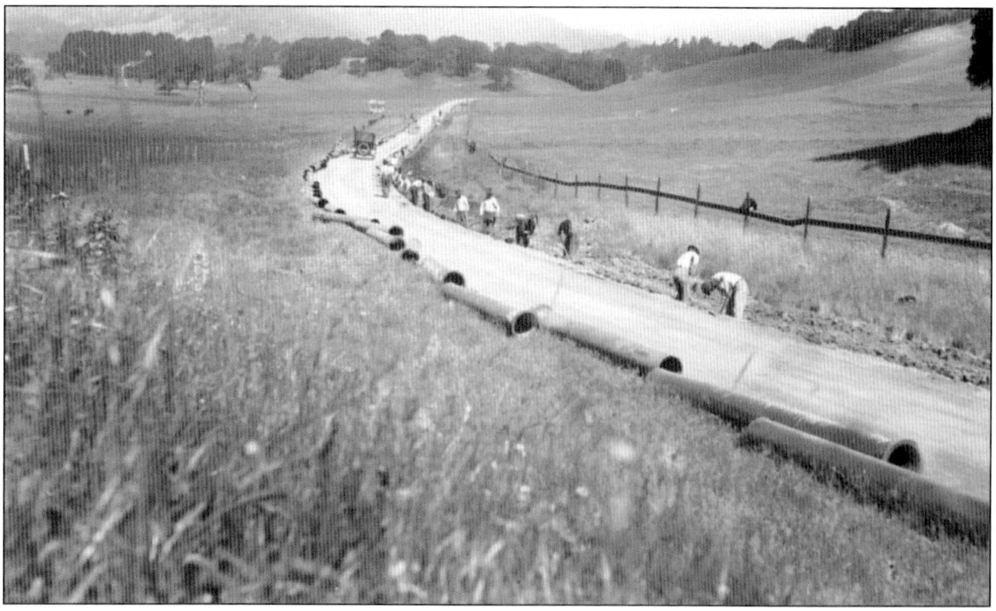

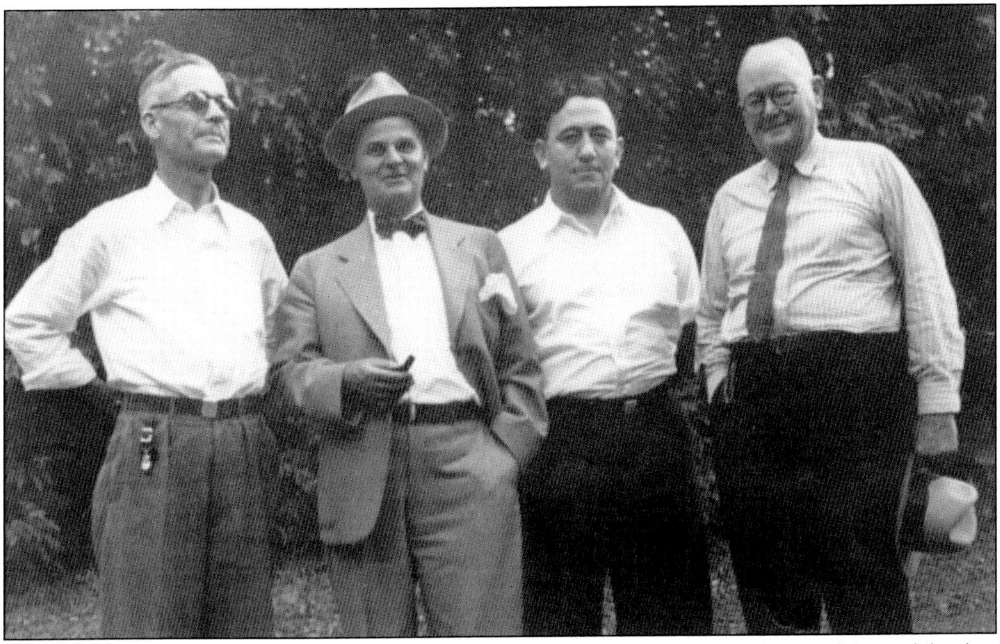

These pictures from the late 1930s show several MMWD senior staff members and four of the five members of the board of directors. The above picture includes, from left to right, James Wraight, general manager James Peters, board secretary B.J. Brusatori, and attorney for the district Robert Duke. Below are the same staff members, plus four members of the board of directors, from left to right, as follows: Duke, Peters, Wraight, Brusatori (kneeling), Carl Symonds (Mill Valley), Les Whitaker (San Rafael), Bert Nickerson (Ross Valley), and Daniel A. McLane (Belvedere). Board member Thomas Kent, the son of William Kent, is not pictured, but he did serve on the board from 1920 until his death in 1959.

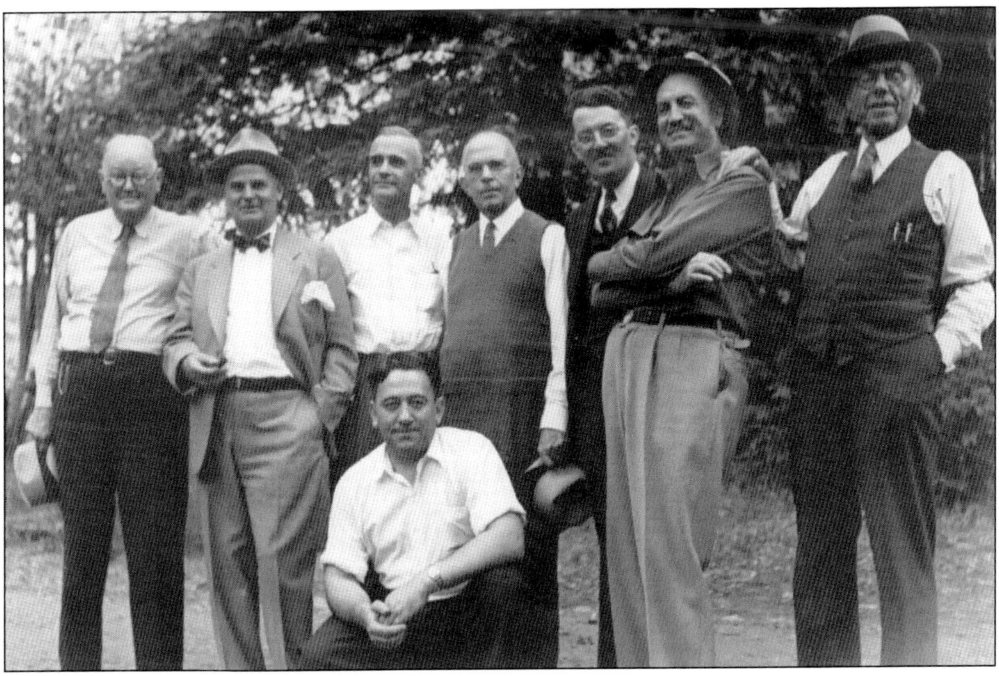

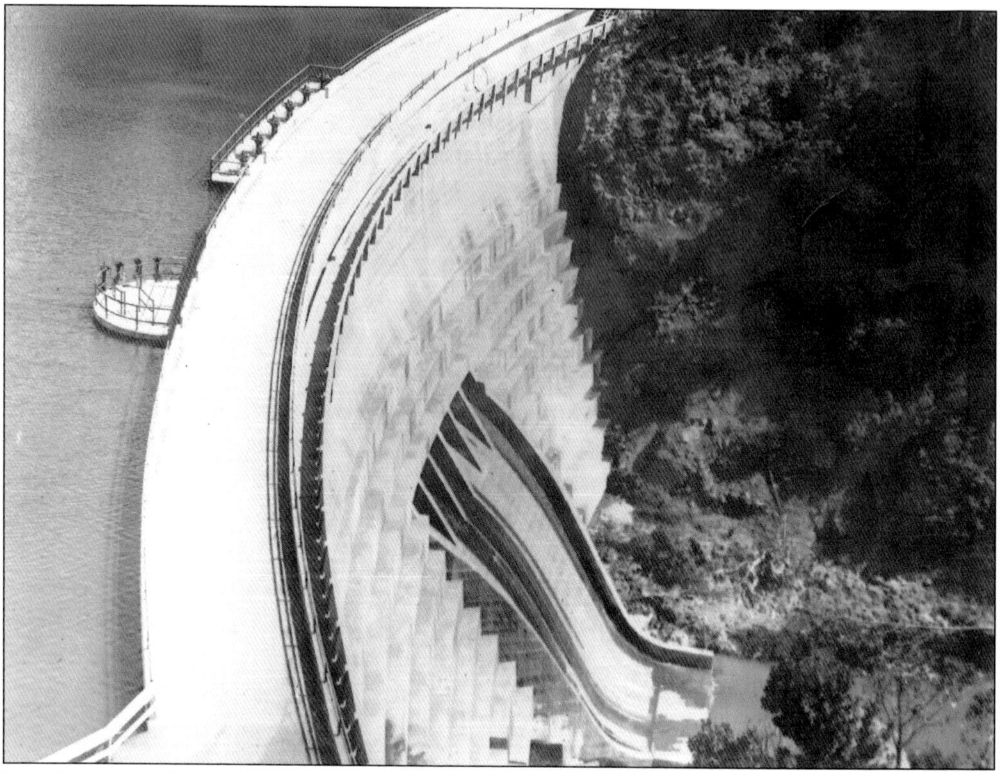

In 1924, C&H offered to expand its water purchase contract with MMWD. To meet that demand, the district increased the height of Alpine Dam by eight feet, as shown here. Under the new contract, C&H agreed to reimburse MMWD for the full cost of raising the dam and entered a contract to make five equal payments, one each year from 1925 to 1930. Albert R. Baker was retained as consulting engineer. The original roadway with circular cutouts in the railing is still visible below the bridge's new roadway and modern railing.

Marin Municipal Water District

SCHEDULE OF RATES

Authorized by Ordinance No. 19 and Effective January 1st, 1929

Marin Municipal Water District
468 FOURTH STREET
SAN RAFAEL, - - CALIFORNIA

In the mid-1800s, as private water companies were allowed to develop, the state required that, so long as supply permitted, water service had to be given to community inhabitants at "reasonable rates," and in the case of fire or other great necessity, free of charge. Community rates were to be set by an elaborately appointed local commission. The system of rate regulation failed and most communities failed to establish rates or even set up the commission. MMWD water rates have always been set by the elected board of directors. This rate schedule shows the rates charged by MMWD in 1929, when a typical monthly bill for a residential customer was $1.50.

Water for General Use

Minimum Charge for Use of--

5/8	inch meter and 300 cu. ft. of water or less	*$1.50*	per month
3/4	inch meter and 300 cu. ft. of water or less	*1.75*	per month
1	inch meter and 300 cu. ft. of water or less	*2.25*	per month
1½	inch meter and 300 cu. ft. of water or less	*3.25*	per month
2	inch meter and 300 cu. ft. of water or less	*5.25*	per month
3	inch meter and 300 cu. ft. of water or less	*8.75*	per month
4	inch meter and 300 cu. ft. of water or less	*13.25*	per month
6	inch meter and 300 cu. ft. of water or less	*25.75*	per month
8	inch meter and 300 cu. ft. of water or less	*40.75*	per month

The charge for water in excess of 300 cu. ft. per month, used under this schedule is as follows:--

From 300 to 5000 cu. ft. Inclusive at *23½c* per 100 cu. ft.
From 5000 to 10000 cu. ft. Inclusive at *22c* per 100 cu. ft.
From 10000 to 15000 cu. ft. Inclusive at *20½c* per 100 cu. ft.
From 15000 to 20000 cu. ft. Inclusive at *19c* per 100 cu. ft.
All over 20000 cu. ft. at *17½c* per 100 cu. ft.

Minimum charge for less than one month service

For 15 days or less---one-half of monthly minimum charge. *For over 15 days*---full monthly minimum charge, based on minimum charge for size of meter in use as shown above.

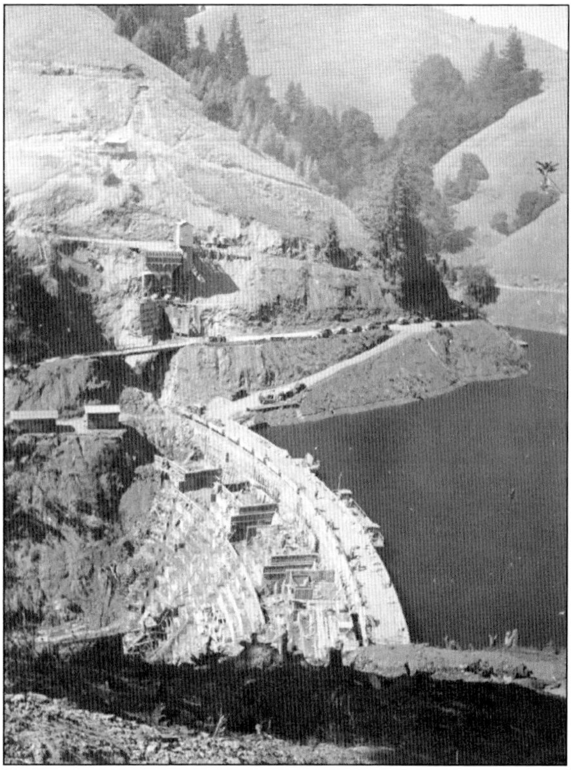

During the Depression years, the Marin population grew and water consumption steadily increased. In 1939, MMWD general manager James Peters advised the board of directors that there was adequate water for the year, but that a second dry year "would severely jeopardize the water supply in the district." Peters recommended that the board take steps as soon as possible to raise Alpine Dam for the second time. Peters consulted with Samuel Morris, the dean of engineering at Stanford University. Various methods of raising the dam were studied and presented to the board for approval. The board reviewed proposals for an arch gravity type and for a gravity section type dam. The board approved the stronger arch gravity type dam, even though the alternative would have been $200,000 cheaper.

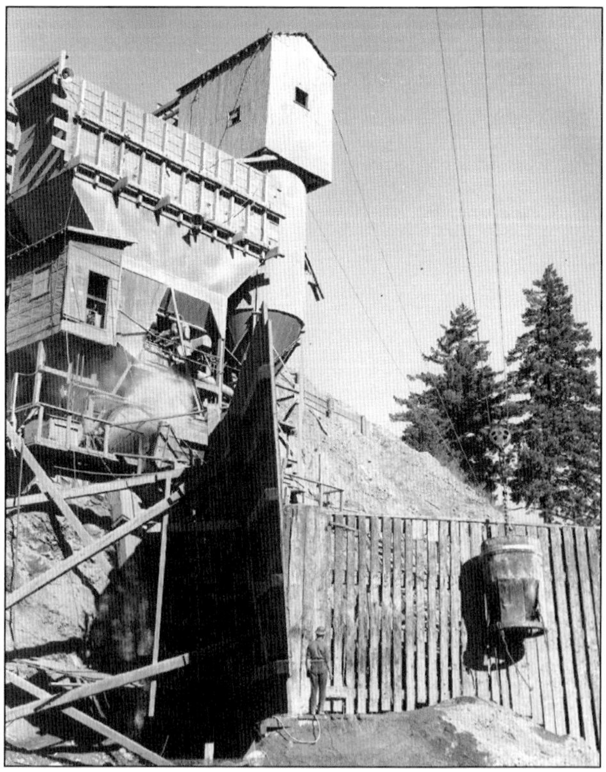

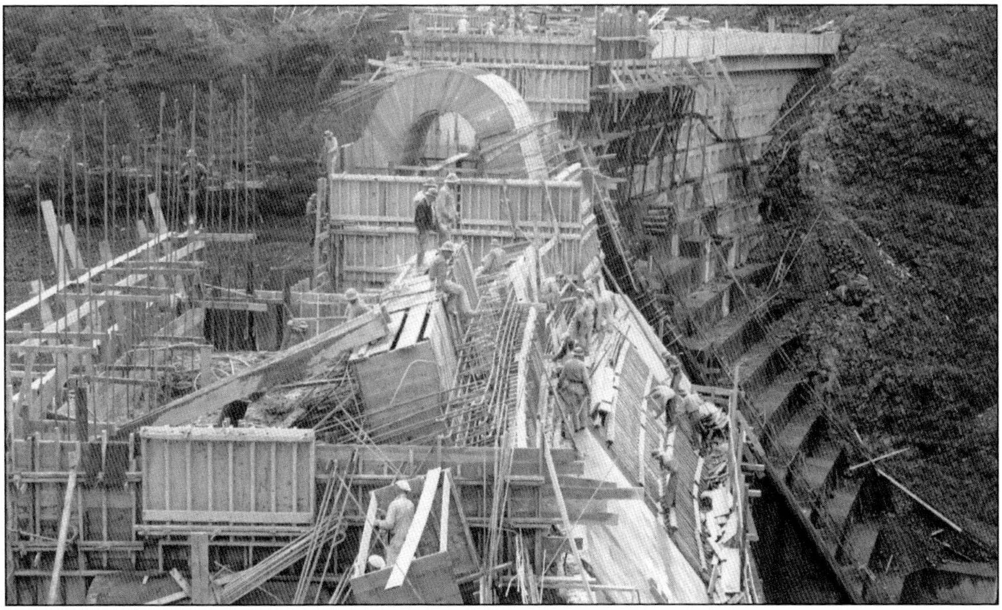

With 30 feet of additional height added, the dam's capacity more than doubled, from one billion gallons in 1918 to 2.4 billion gallons in 1941. Concrete for the new addition was first poured one week after the Pearl Harbor attack. Timing was critical because available manpower rapidly diminished due to the war. The contract for construction of the dam was awarded to Engineers Limited on their low bid of $469,397. In 1941, the Depression was still in view, and as the site was being cleared, the wood was saved for distribution to the needy. The wood needed to be cut and split and readied for distribution, and the Presbyterian Orphanage and Farm (now called Sunny Hills Services) in San Anselmo agreed to participate in that effort. (Courtesy of the Anne T. Kent California Room, Marin County Free Library.)

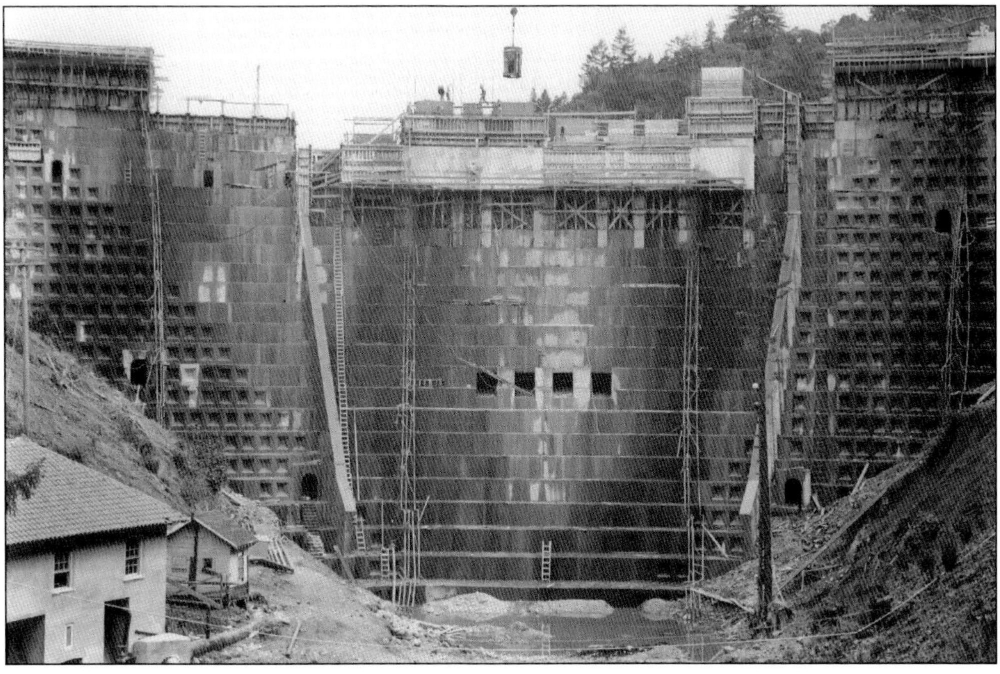

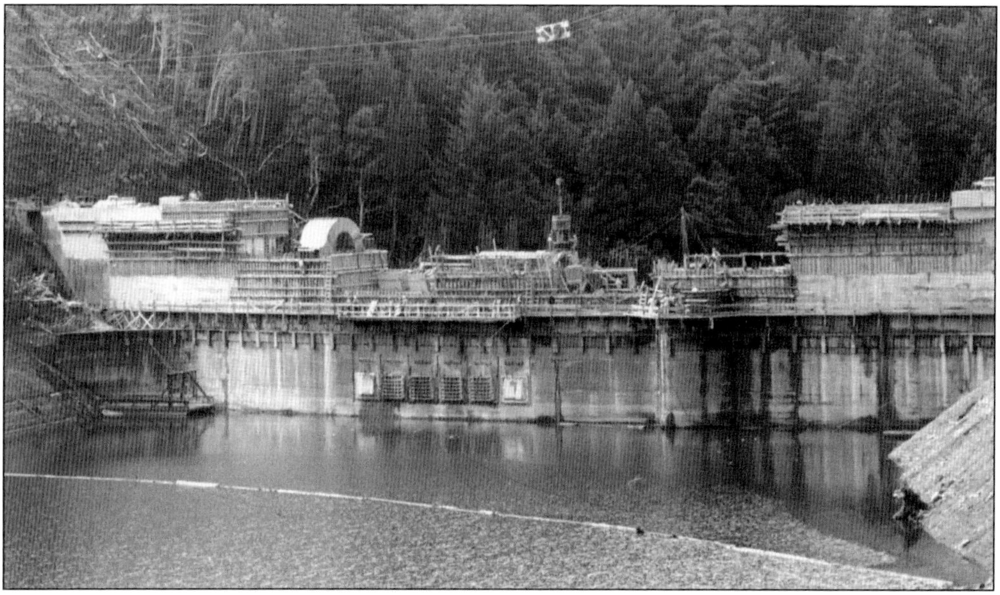

Raising the height of the dam required an increase in its total cross-section and thickness, making it a much more complex engineering task than the 1924 raising. A 40-foot-thick slab of concrete was added at the base of the dam, extending from the rock foundation of the old dam on an 8-to-1 slope to the crest of the new dam. Approximately 40,000 yards of concrete and 300 tons of reinforcing steel were used. (Courtesy of the Anne T. Kent California Room, Marin County Free Library.)

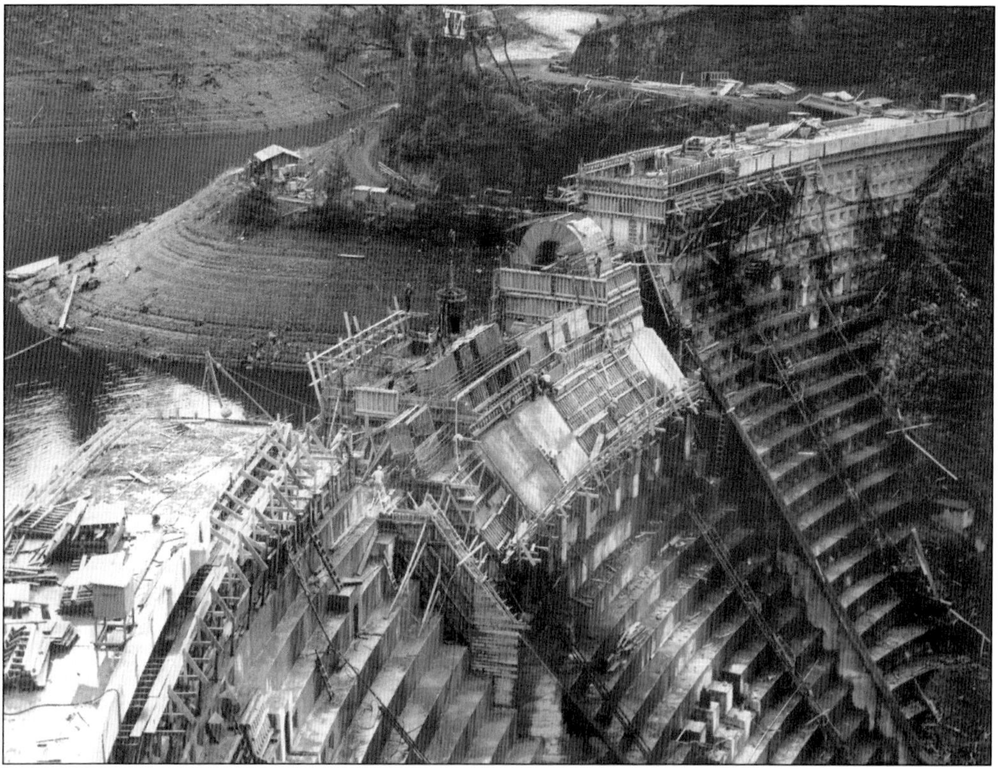

Work on the raised dam was completed on June 16, 1942, for $728,640. The contractor's portion went over budget by almost $200,000 due to excavation required to find bedrock to support the added weight. The new dam had a water surface elevation of 646 feet behind a 135-foot-high concrete structure that was 115 feet thick at the base and 21 feet wide along the 530-foot crest.

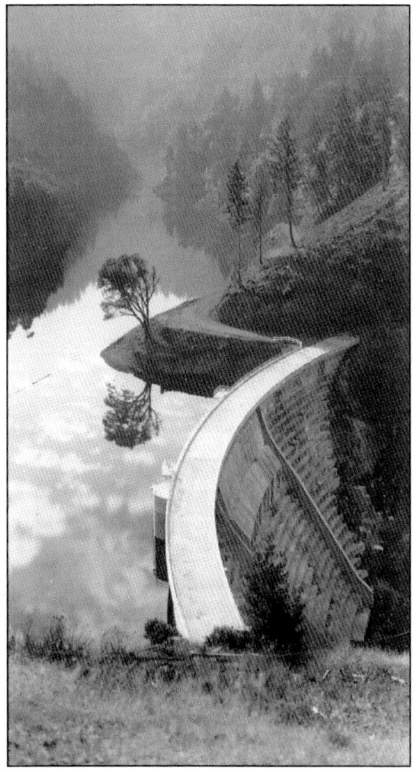

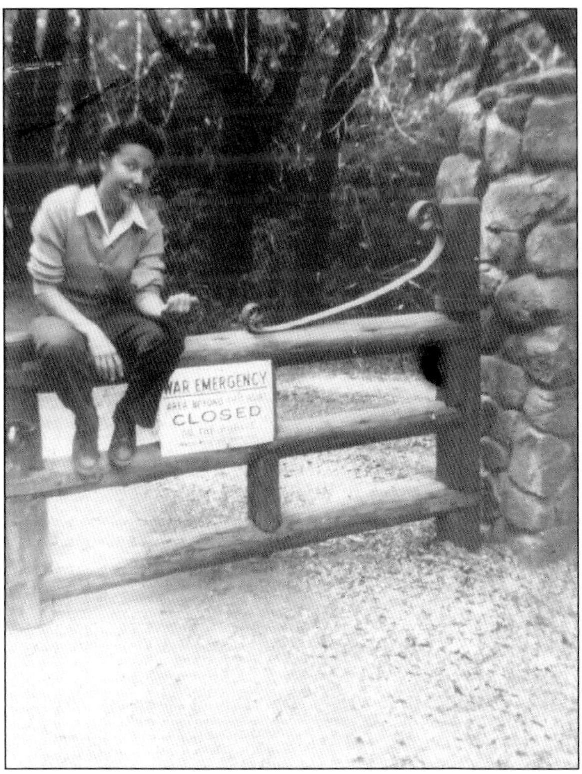

With the outbreak of World War II on December 7, 1941, the dam and other district facilities were placed under guard by district employees. Two days later, the State Guard arrived and remained on duty at Alpine Dam until May 2, 1943. The woman in the photograph is at the gate to Phoenix Lake. The sign reads "War Emergency—Area Beyond This Point Closed to the Public."

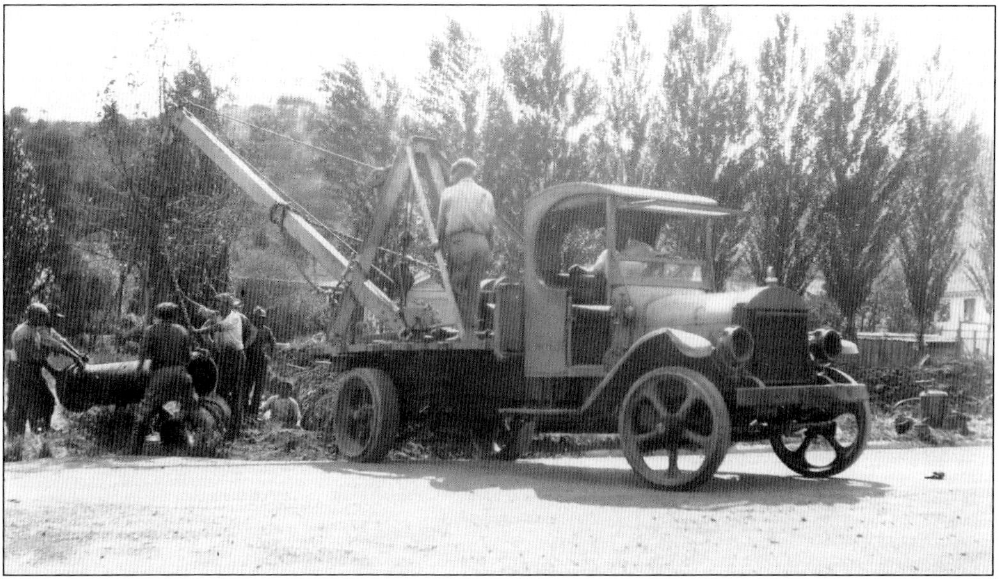

The laying and repairing of pipe is an almost continuous job for MMWD. In the above photograph, workmen lower pipe into a ditch using a flatbed truck modified with a crane and motorized winch. This method, used in the late 1920s, is similar to methods of today, but the modern equipment has greater safety features.

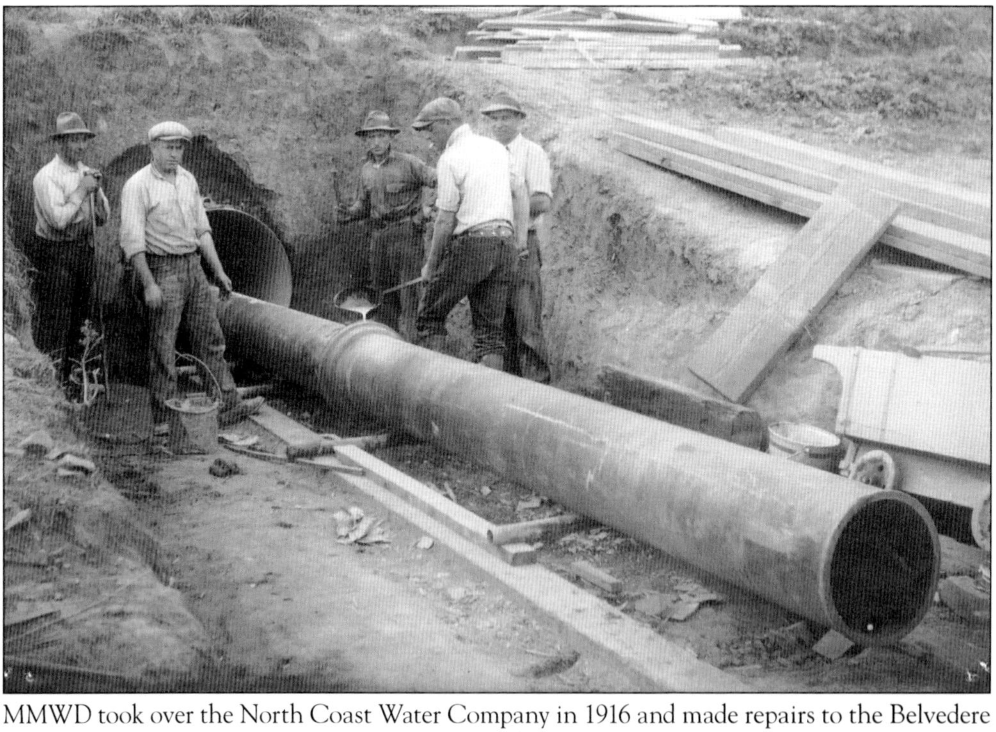

MMWD took over the North Coast Water Company in 1916 and made repairs to the Belvedere Dam in Mill Valley on several occasions. In this photograph, a 42-inch pipe was jacked through the north embankment of the dam and a smaller, 14-inch cast-iron outlet pipe was installed inside the larger pipe. Pictured from left to right are MMWD employees Charlie Zaro, Joe Beltramini, foreman Joe Merla, Charlie Merla, and Emilio Gerrito.

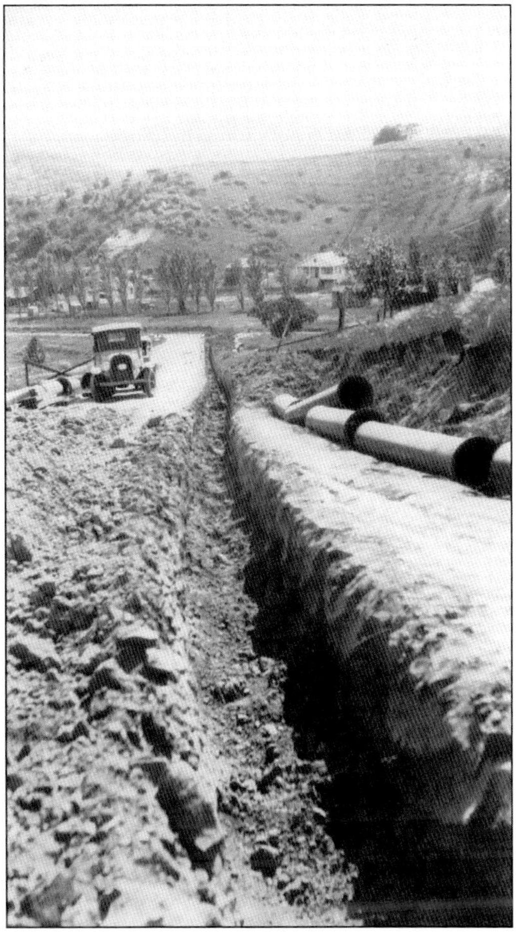

After World War I, the War Department decided to build a West Coast bomber base in Marin on a section of flat land beside San Pablo Bay. President Hoover signed a bill providing more than $1 million for the airfield. Agreement was reached with MMWD, and the district began laying a pipeline from the 200,000 gallon Hawthorne Hills Tank off Camino De Herrera in San Anselmo through Sleepy Hollow Ranch, over the ridge near Freitas Ranch, and onto St. Vincent's School for Boys to a holding tank near the proposed airfield.

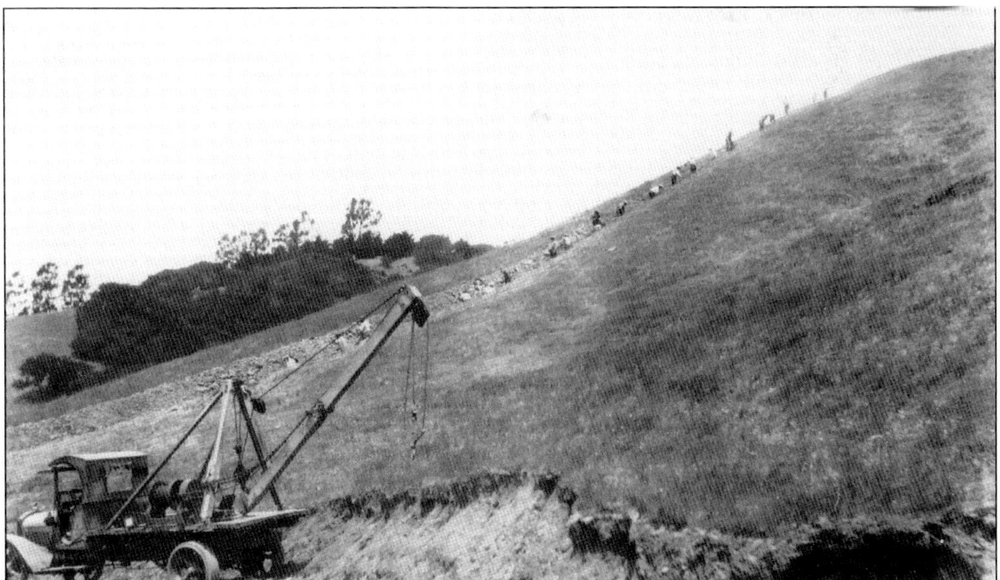

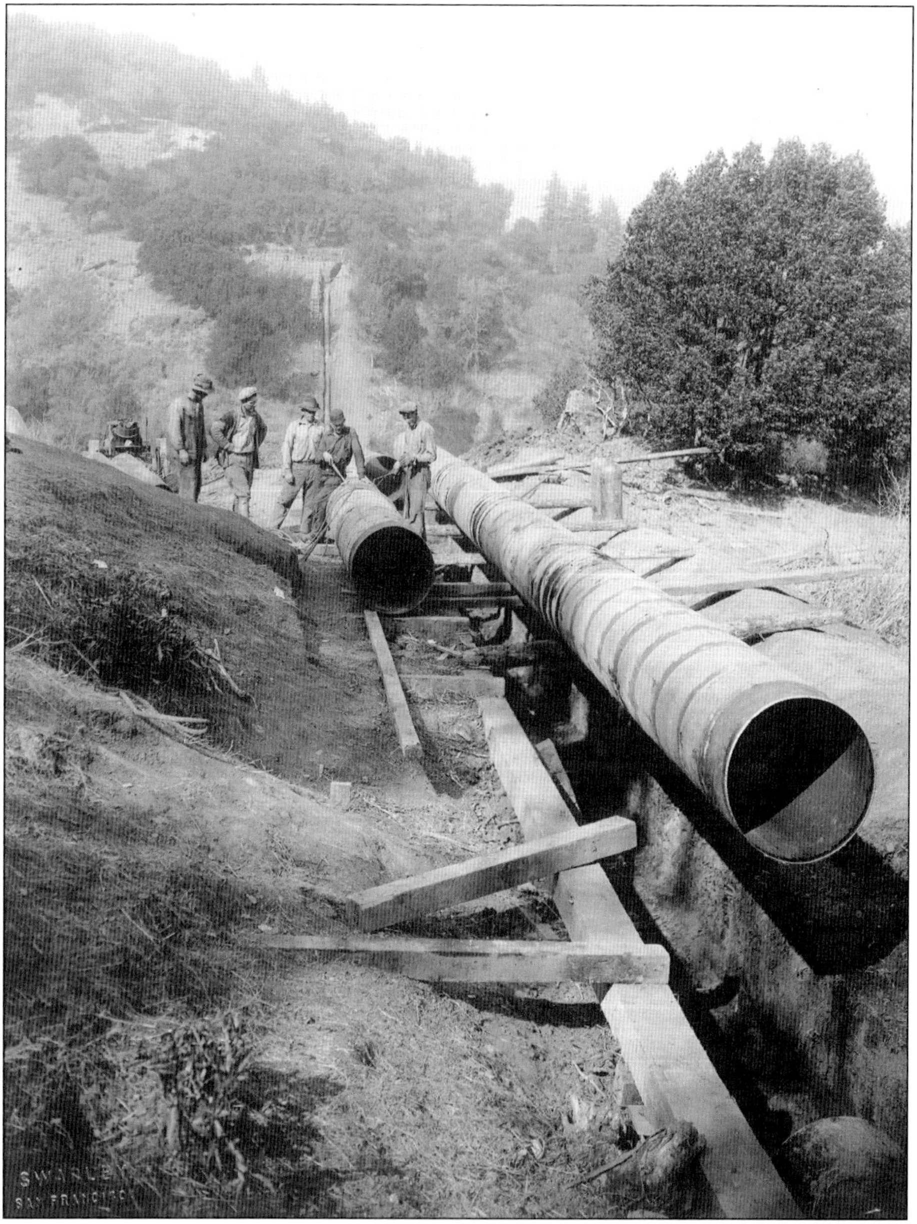

The Marin Municipal Water District has more than 900 miles of transmission and distribution pipe throughout its service area. Much of the pipe installed in the past 40 years has a life span of more than 100 years. The district's pipeline replacement program is designed to anticipate areas of pipeline failure before they actually occur. The district continually calculates the remaining service life of pipelines based on the type and age of the pipe, repair history, and the soil makeup in the area. From these evaluations, candidates for the annual pipeline replacement program are identified. Today's plastic pipeline technology suggests that pipes will have an indefinite life span. In this 1926 photograph, MMWD employees install a welded steel pipe water main between the town of Fairfax and Phoenix Lake. The beautiful hills of Marin County have been a challenge to the construction of distribution water mains, with pipelines often laid down steep slopes at difficult angles. (Photograph by Wesley W. Swadley.)

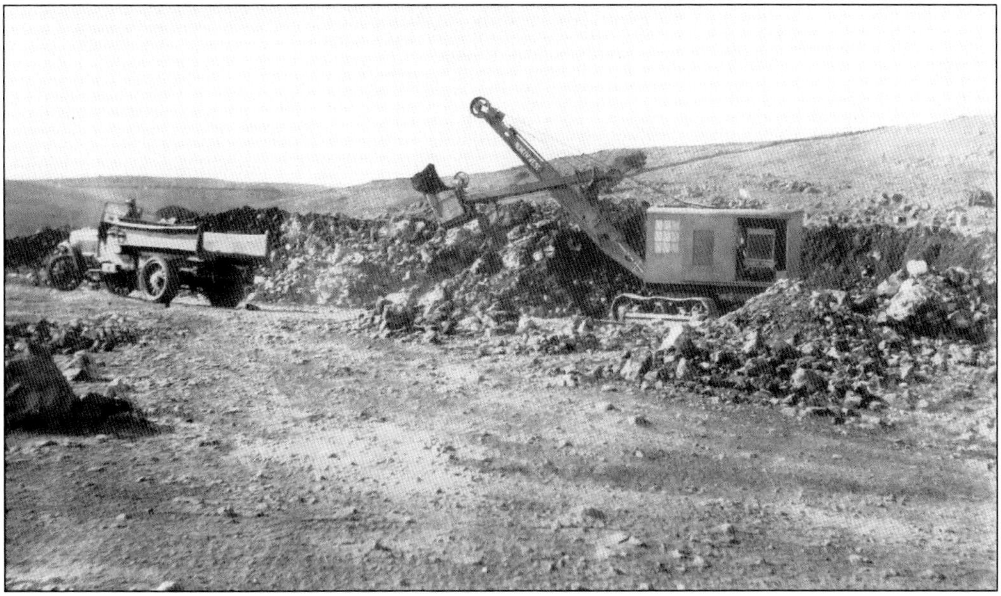

In 1927, a Belvedere Island tank received considerable attention, described in the September 1927 MMWD customer newsletter as a detriment to the "scenic advantages of the neighborhood." This new 1.5-million-gallon steel tank was built in the hills north of Tiburon to replace it. On land previously part of the Deffebach Estate, the new tank, and a smaller 200,000-gallon tank above it, allowed the Belvedere tank to be removed. Today, MMWD uses placement, color, and landscaping to minimize the aesthetic disturbance of necessary tanks.

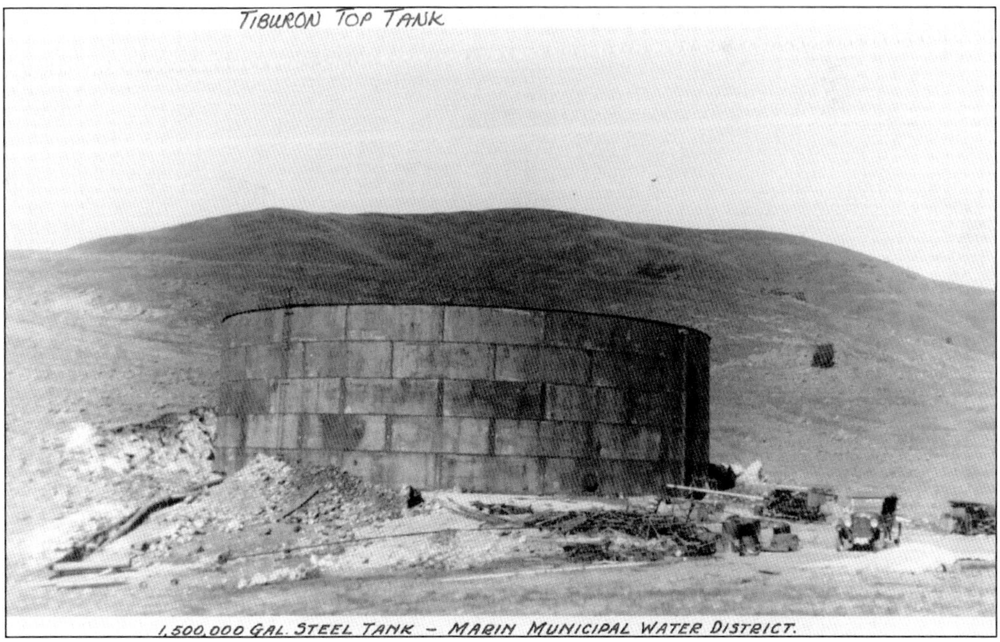

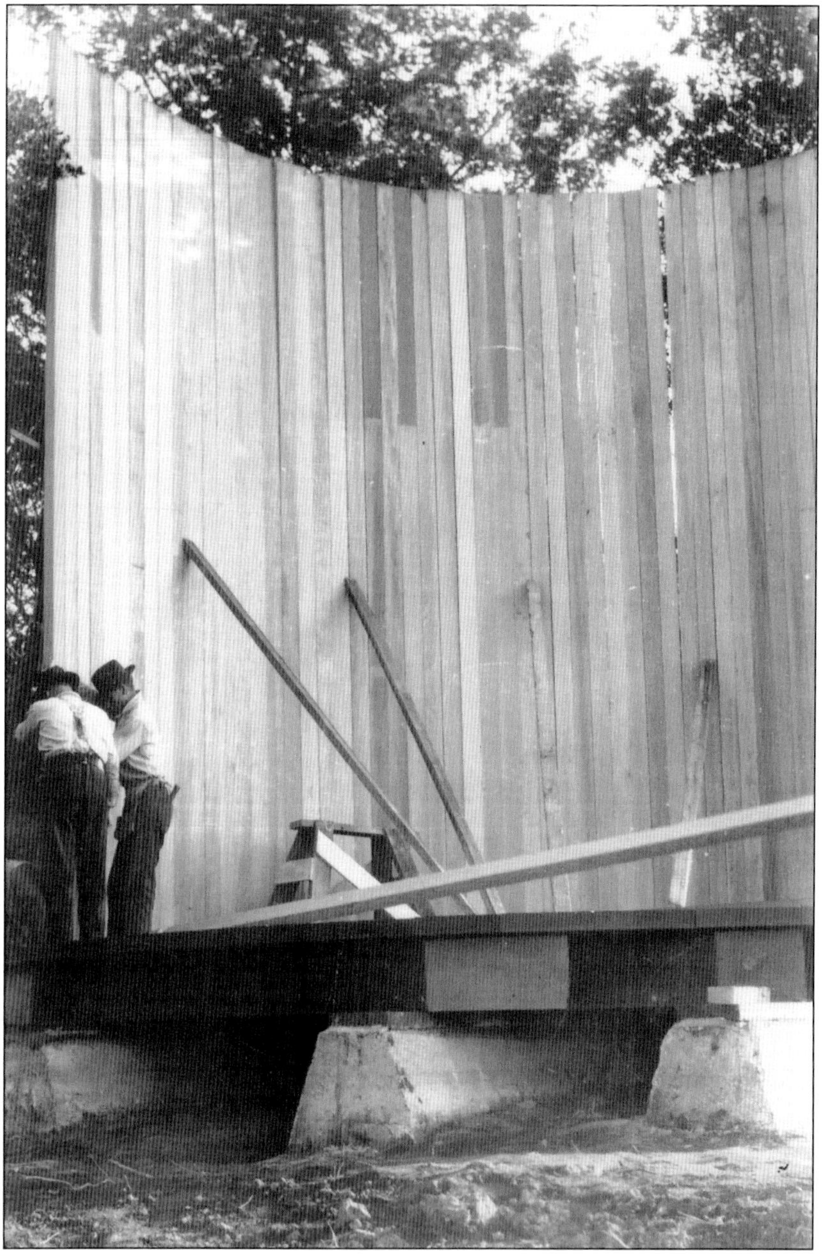

Redwood tanks have historically been an important part of the MMWD water system. In this 1935 photograph, MMWD employees assemble a redwood water storage tank. Each of the redwood staves had to be carefully placed and then collectively banded together with steel rods. Throughout the 19th and the early 20th centuries, redwood was often used as a building material. In 1873, the Marin Water Company used a redwood flume as part of Lagunitas Dam and Reservoir. The flume, still in place today, although rebuilt, was used as piping to start the flow of water from the reservoir on its journey to downtown San Rafael. Redwood tanks were an economical and accepted method of water storage. Today, recognizing that redwood tanks require greater maintenance and present seismic instability, MMWD is in the process of phasing out all redwood tanks and replacing them with steel tanks.

These aerial views show the location of the old Lagunitas Dairy Tract and the same site after Bon Tempe Dam and Reservoir were constructed. In 1868, the three Bautunpi brothers leased 1,180 acres of land along Lagunitas Creek and for many years ran the Lagunitas Dairy Tract, a very successful dairy ranch. By 1874, the ranch was supplying the US Army, San Quentin, and the people of San Rafael with thousands of pounds of butter annually. The local newspapers Americanized the "Bautunpi" name as "Bontempi," which later evolved to "Bon Tempe." The land was purchased by the Marin County Water Company in 1878, taken by MMWD in 1916, and eventually used to build the Bon Tempe Dam and Reservoir in 1948.

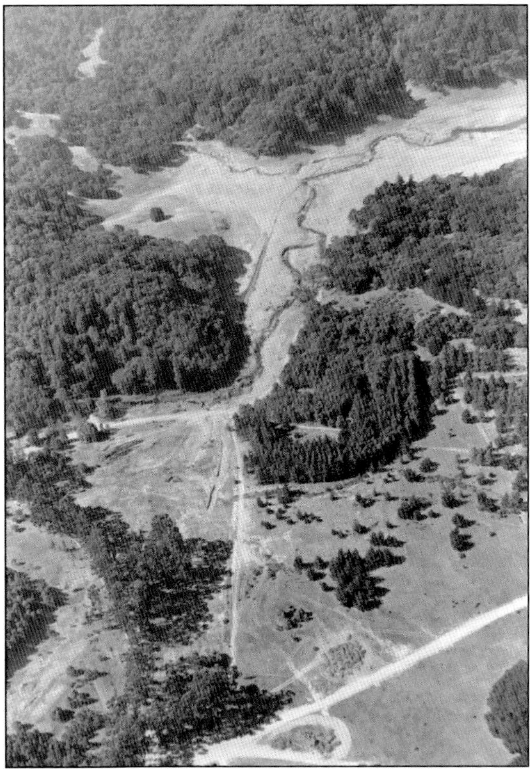

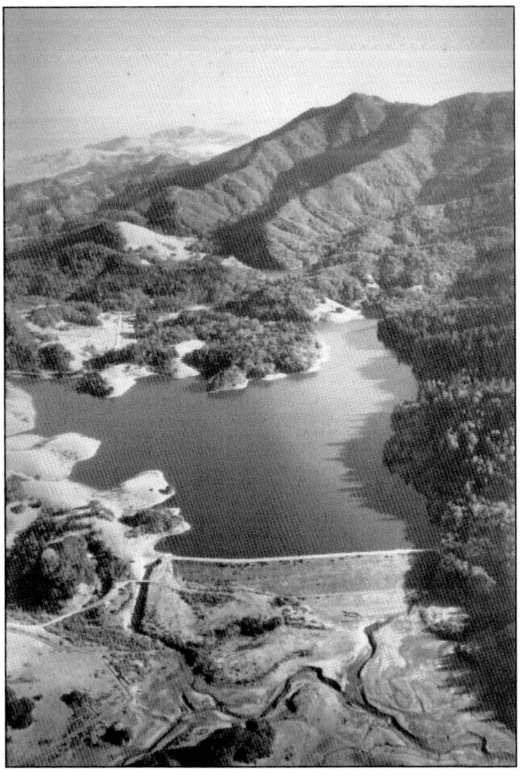

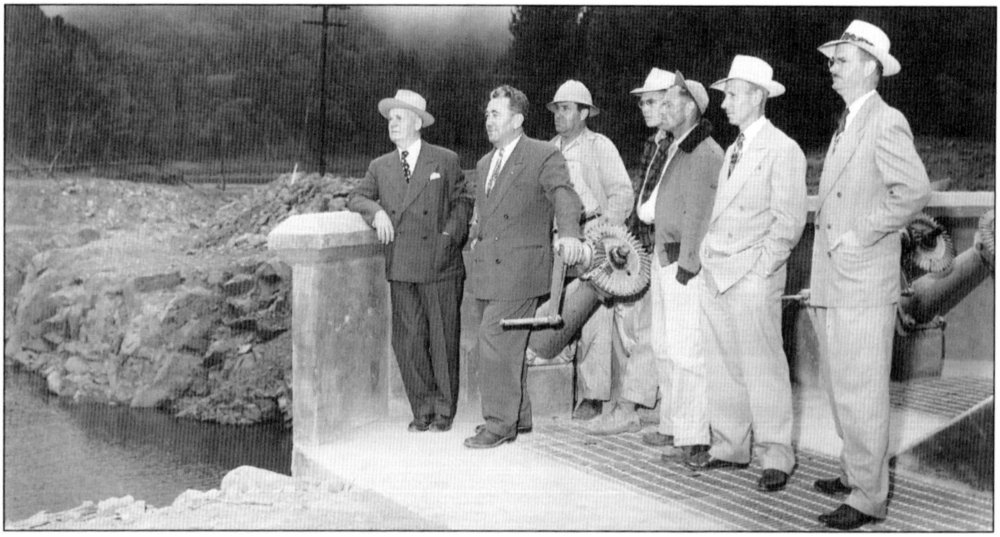

This concrete inlet structure with seven control gates was built in 1951 at the east end of Bon Tempe Reservoir. Water is taken through the gates into a pipe through the Bon Tempe Tunnel. Pictured from left to right are general manager James Peters, secretary-treasurer B.J. Brusatori, engineer-in-charge Donald Costello, two unidentified men, assistant general manager William R. Seeger (who later served as general manager), and Rockwell representative G. Bowman. (Photograph by George Wheeler.)

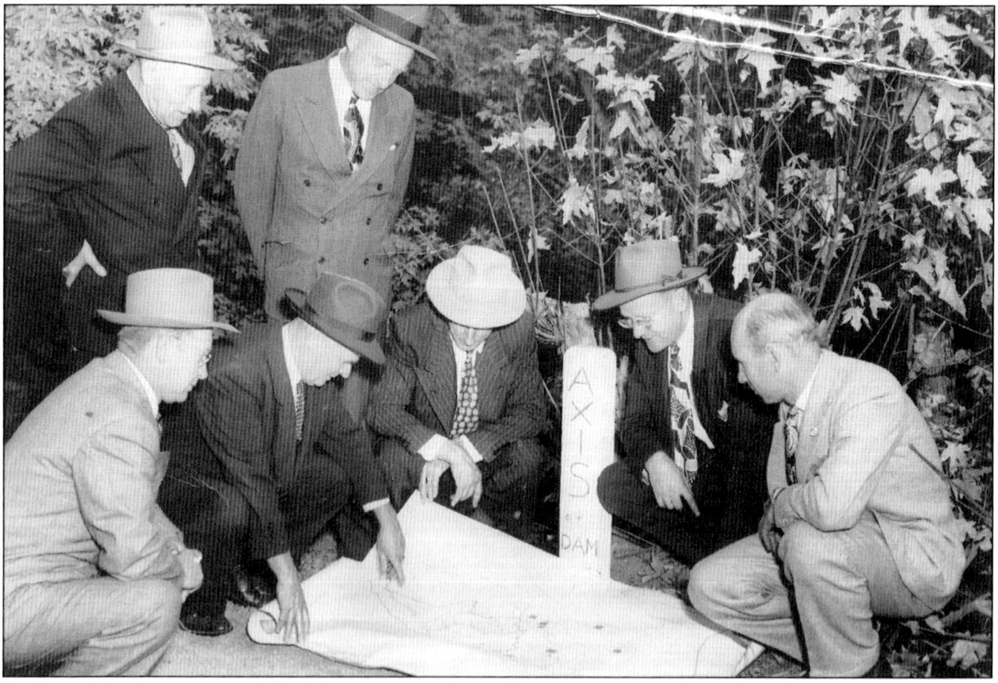

Assistant general manager William Seeger (right, with no hat), general manager James Peters (kneeling, with his hands on the map), and others visit the site of the proposed Peters Dam in 1953. The dam was named for James Peters, the former MMWD general manager who spent 42 years with the district. The reservoir was named Kent Lake after Thomas T. Kent, who served on the board of directors for 39 years.

Herb K. Brainerd, the water district's resident engineer, works on the design of Peters Dam in 1953. Like many early MMWD engineers, he started his career working for Michael O'Shaughnessy in San Francisco. In 1921, Brainerd was hired by San Rafael mayor George Alpers (who later served as an MMWD board member) to serve as city manager.

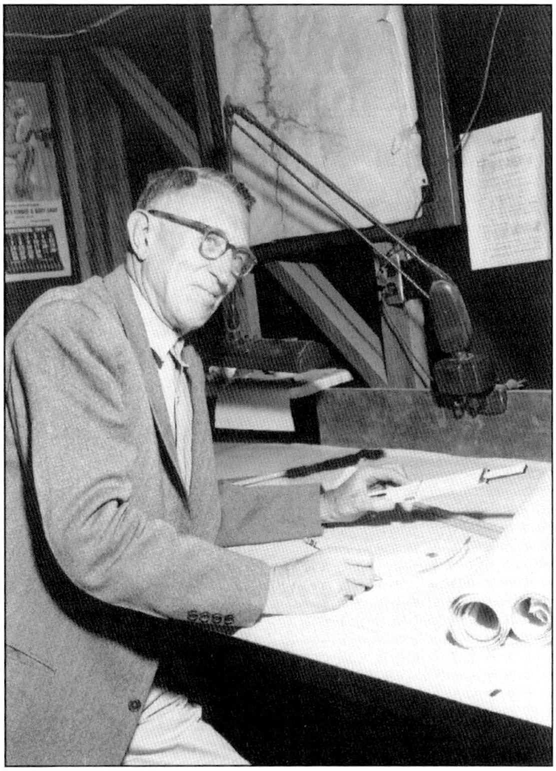

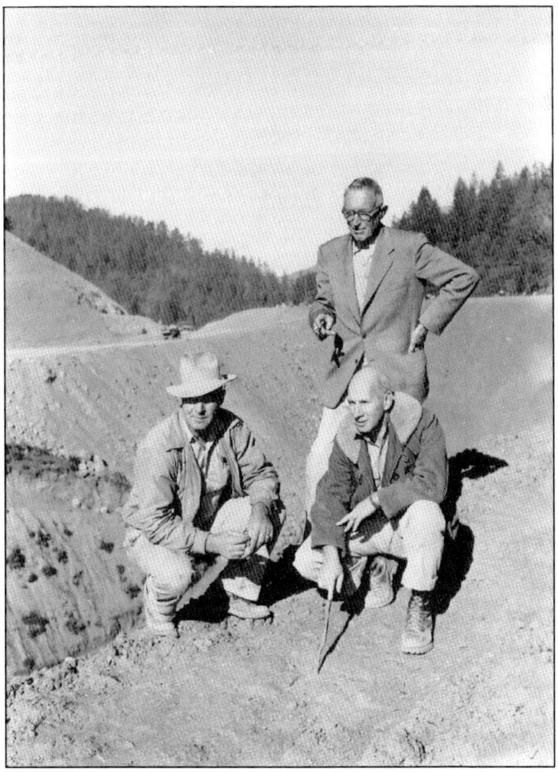

Resident engineer Herb K. Brainerd (standing) points something out as Earl Bernard, superintendent for the John Delphia construction firm (left), and William Seeger kneel down on a 1953 site visit to Peters Dam.

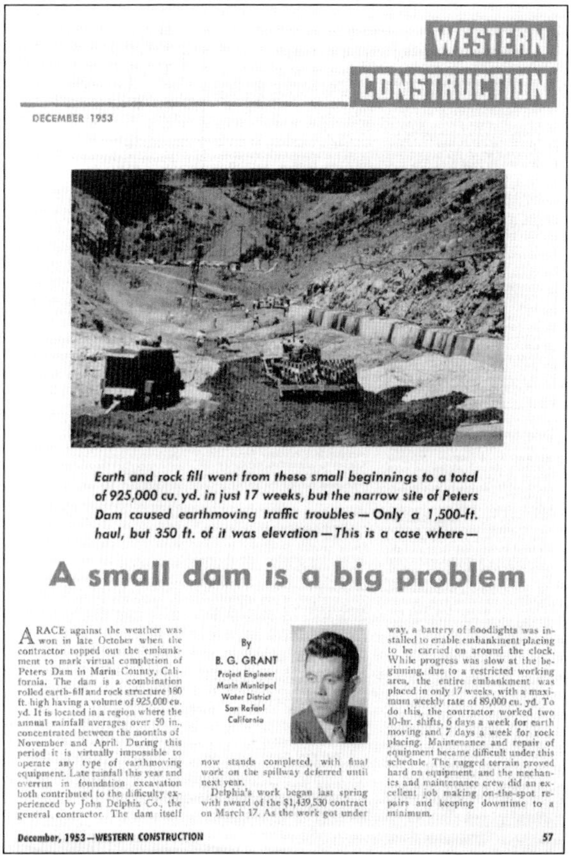

WESTERN CONSTRUCTION

DECEMBER 1953

Earth and rock fill went from these small beginnings to a total of 925,000 cu. yd. in just 17 weeks, but the narrow site of Peters Dam caused earthmoving traffic troubles — Only a 1,500-ft. haul, but 350 ft. of it was elevation — This is a case where —

A small dam is a big problem

By
B. G. GRANT
Project Engineer
Marin Municipal Water District
San Rafael
California

A RACE against the weather was won in late October when the contractor topped out the embankment to mark virtual completion of Peters Dam in Marin County, California. The dam is a combination rolled earth-fill and rock structure 180 ft. high having a volume of 925,000 cu. yd. It is located in a region where the annual rainfall averages over 50 in., concentrated between the months of November and April. During this period it is virtually impossible to operate any type of earthmoving equipment. Late rainfall this year and overrun in foundation excavation both contributed to the difficulty experienced by John Delphia Co., the general contractor. The dam itself now stands completed, with final work on the spillway deferred until next year.

Delphia's work began last spring with award of the $1,439,530 contract on March 17. As the work got under way, a battery of floodlights was installed to enable embankment placing to be carried on around the clock. While progress was slow at the beginning, due to a restricted working area, the entire embankment was placed in only 17 weeks, with a maximum weekly rate of 89,000 cu. yd. To do this, the contractor worked two 10-hr. shifts, 6 days a week for earth moving and 7 days a week for rock placing. Maintenance and repair of equipment became difficult under this schedule. The rugged terrain proved hard on equipment and the mechanics and maintenance crew did an excellent job making on-the-spot repairs and keeping downtime to a minimum.

December, 1953—WESTERN CONSTRUCTION 57

In 1954, MMWD completed construction of Peters Dam. The dam and its method of construction received attention from the professional engineering community. Bill Grant, who served as a project engineer with resident engineer Herb K. Brainerd, wrote an article for the professional journal, *Western Construction*, in December 1953, describing the problems they encountered building this dam on a narrow gorge on Lagunitas Creek.

As construction continued throughout 1953, earth-moving became a problem because the working site of the dam was so narrow. This flash flooding occurred while an outlet pipe was being installed in April 1953 and delayed the contractor's schedule.

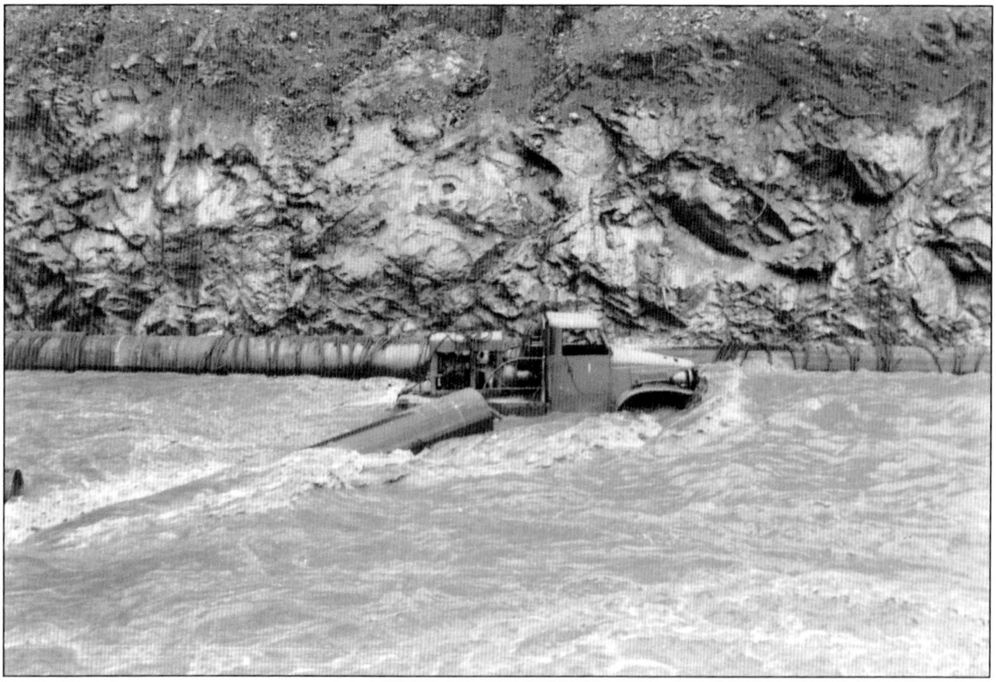

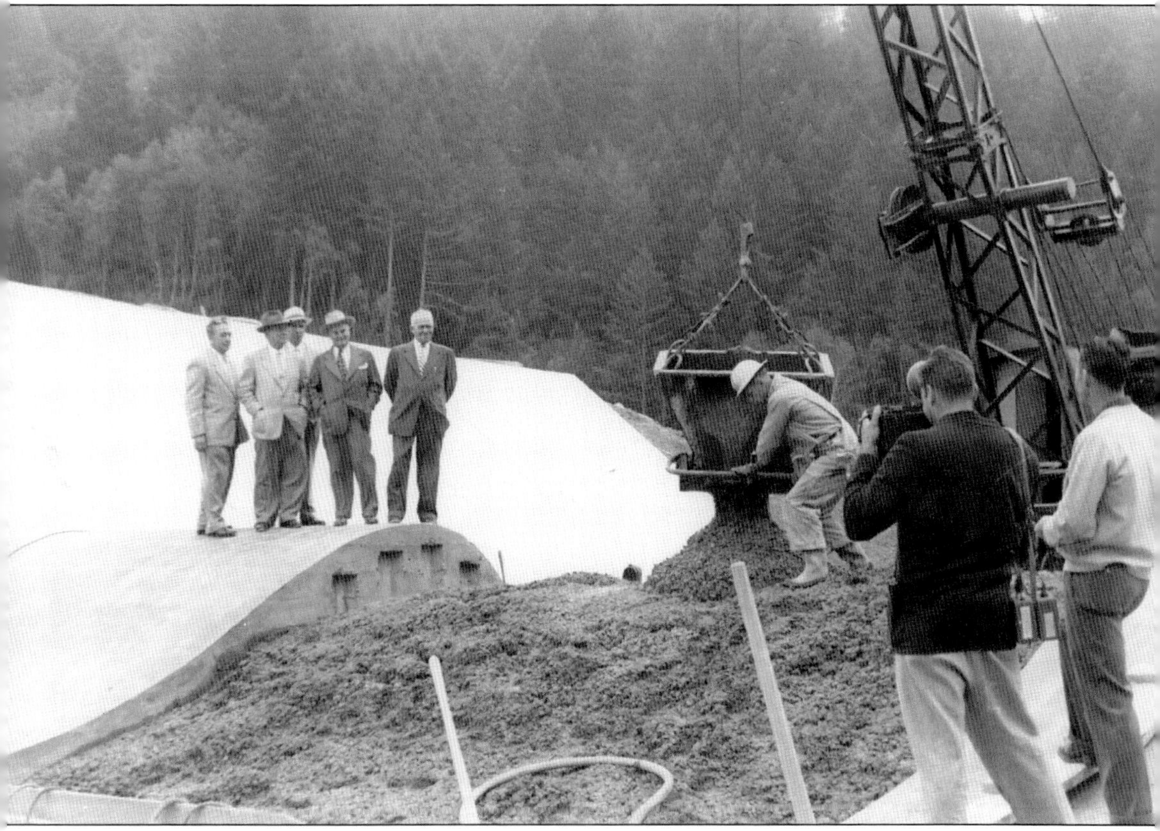

Posing for a photograph of the "last pour" at the spillway of Peters Dam on September 21, 1954 are, from left to right, secretary B.J. Brusatori, director William C. Herup, assistant general manager and assistant engineer William R. Seeger, general manager and chief engineer James Peters, and board president Thomas T. Kent. Peters Dam is a combination rolled earth-fill and rock structure that is 180 feet high. To meet a rigorous construction schedule, the contractor, John Delphia Company, under battery-charged floodlights, worked two 10-hour shifts on a six-day workweek. Over a 17-week period, the contractor moved 89,000 cubic yards of material per week. The rugged terrain around the construction site made construction difficult, causing equipment to require frequent repairs. The district received one problematic report that showed faulting and fracturing in the region, but in the end, the soil condition proved to be sufficiently watertight.

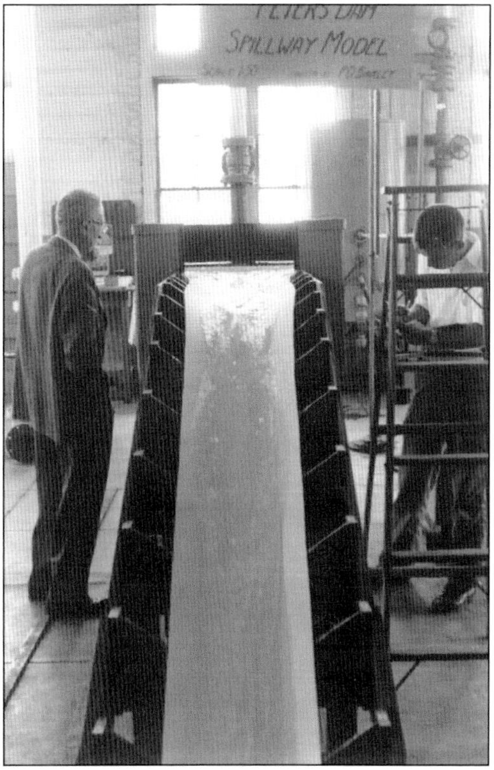

Working with Stanford University engineers, a model for the Peters Dam spillway was built at Stanford University in 1953 (left). Located on the left side of the dam, the spillway consists of a reinforced concrete-lined chute channel with an ogee overflow weir. Its crest is 124 feet long, converging over a 200-foot section into a 30-foot trapezoidal channel. The spillway terminates with a bucket-shaped structure at the bottom, for dissipation of energy. In 1982, a new spillway was designed for Peters Dam in order to raise the dam 45 feet. A model was built and tested by district staff at the fish rearing ponds below Lagunitas Dam (below).

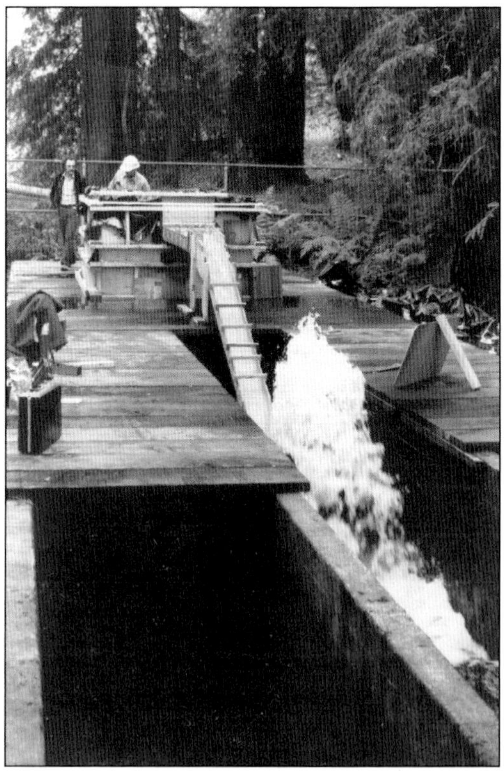

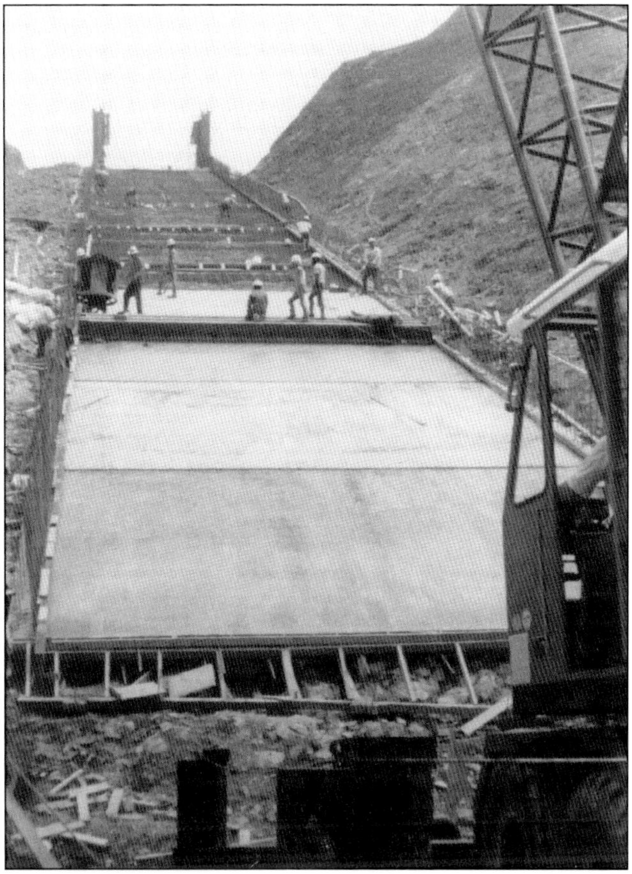

The raising of Peters Dam in 1982 required construction of a new spillway, seen at right during construction. Below, an aerial view of Peters Dam and the reservoir shows the 1982 spillway on the right side of the dam and the 1953 spillway on the left side, with Kent Lake behind it. Peters Dam was originally named Carson Dam, but the board changed the name to Peters Dam, after James Peters, the MMWD general manager from 1935 to 1954.

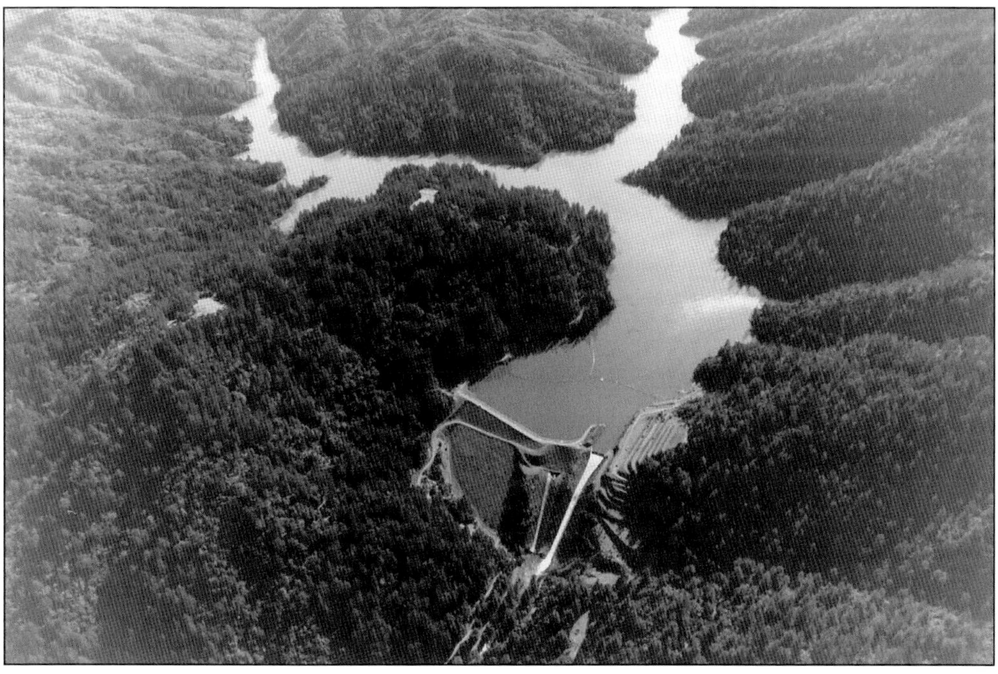

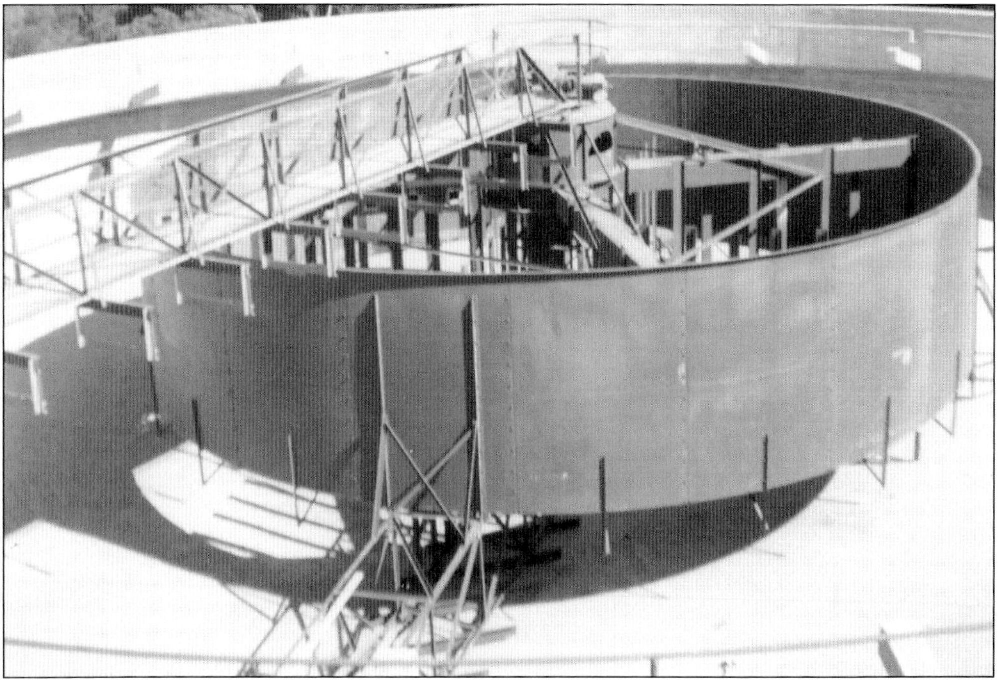

At the August 29, 1956, board meeting, general manager William Seeger presented a proposal for improvements to the MMWD system. The district was faced with the postwar population boom and constant pressure for more water. Among the list of improvements were the Nicasio Dam, the Corte Madera office building, and the Bon Tempe and San Geronimo Treatment Plants. The board sent a $12.6 million bond measure to the voters, which they approved, 33,385 votes to 7,741 votes. All the improvements were completed, and the Bon Tempe Treatment Plant (pictured here) started operating in 1959. The minutes of that August meeting show that the board received "a petition bearing seven signatures and a letter . . . in protest of the proposed action of erecting a large water storage dam and reservoir in the Nicasio Valley."

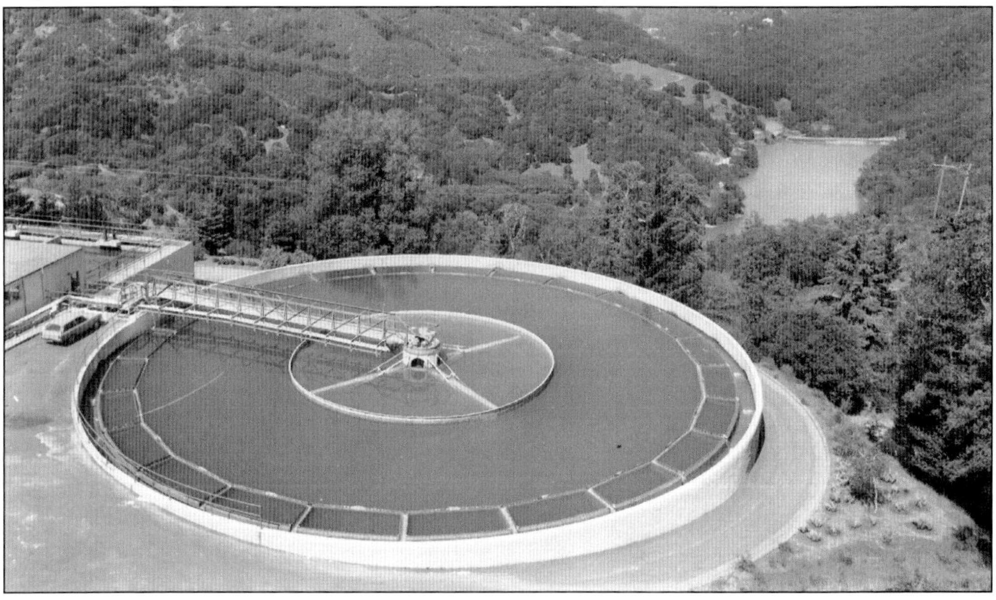

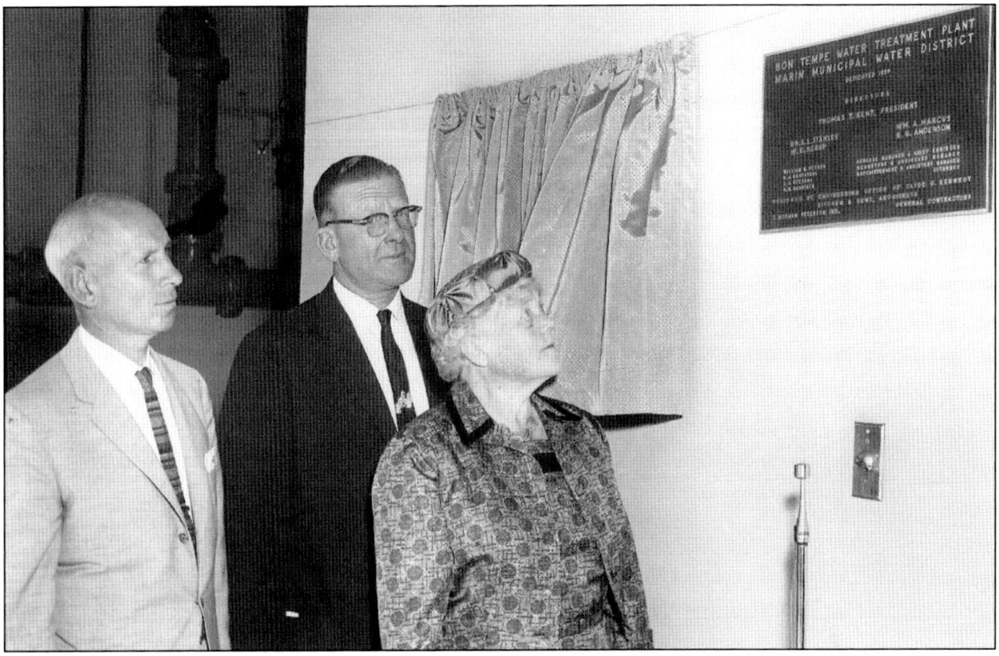

At the July 5, 1959 dedication of the Bon Tempe Water Treatment Plant (from left to right), general manager William Seeger, board member Robert Anderson, and Anne Thompson Kent, widow of director Thomas T. Kent, study a new plaque commemorating the day. Mr. Kent died on March 24, 1959, just a few months before this dedication ceremony. (Photograph by Jim Kean Photo.)

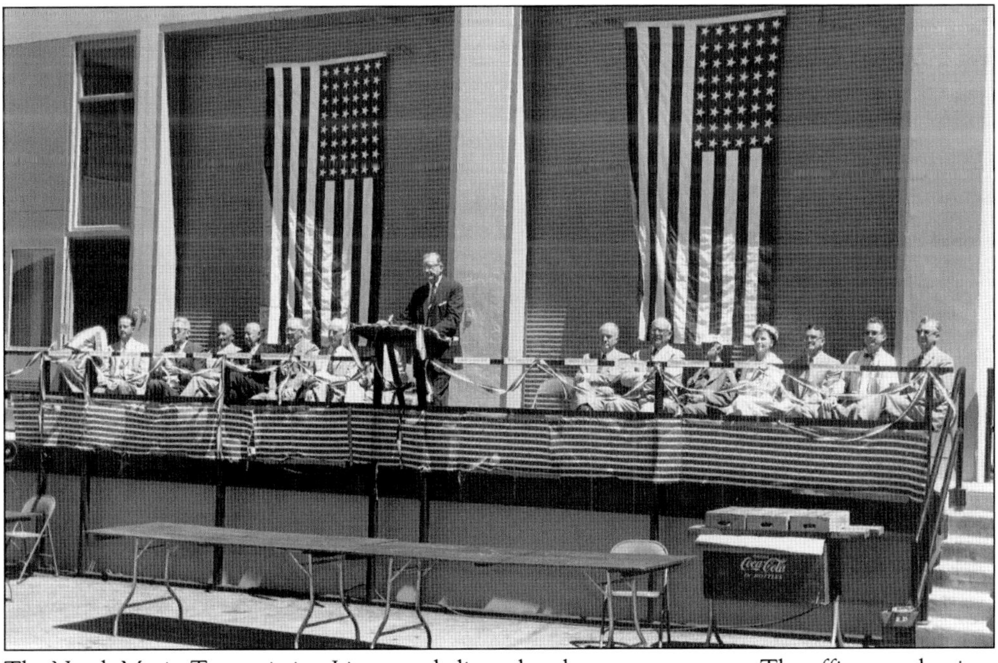

The North Marin Transmission Line was dedicated at the same ceremony. The officers at the time were as follows: board president Robert Anderson, general manager/chief engineer William Seeger, secretary/assistant manager B.J. Brusatori, general superintendent/assistant manager Louis Stefani, attorney Samuel Gardiner, and auditor George Franzman. (Photograph by Jim Kean Photo.)

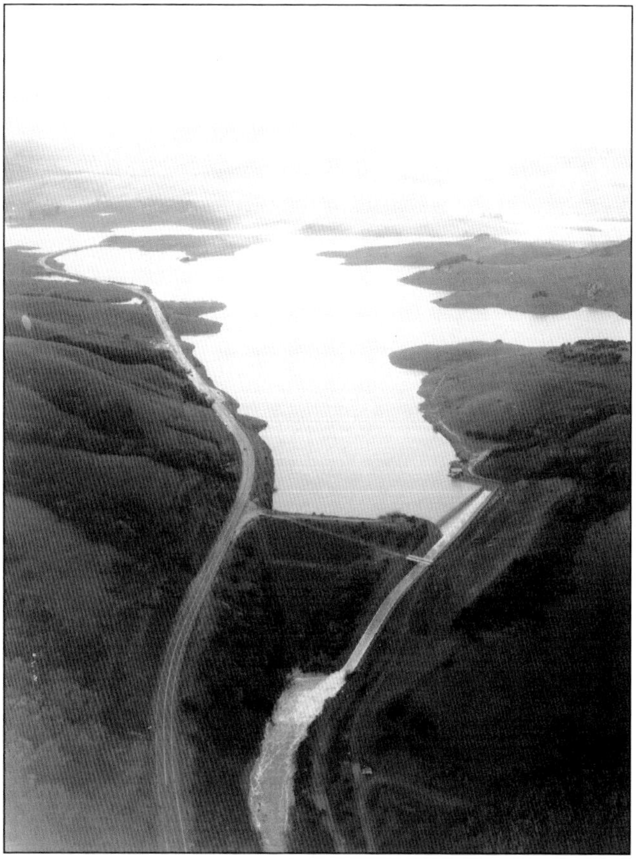

Completed in 1960, the Nicasio Dam and Reservoir are located in the Nicasio Valley of West Marin on Nicasio and Halleck Creeks. It is one of MMWD's sources of water supply located outside the Mount Tamalpais watershed. In 1993, the board of directors renamed the dam after former general manager William Seeger. Created by the 115-foot-high, 400-foot-long earthen dam, the reservoir surface area covers 845 acres and holds 22,430 acre feet or 7.4 billion gallons of water—about 28 percent of the district's total reservoir capacity. The water from this reservoir and Kent Lake is taken to the San Geronimo Treatment Plant for treatment and distribution. The water from these two sources provides most of the drinking water for the northern and central populations of the district's service area. (Left, photograph by Aero Photographers.)

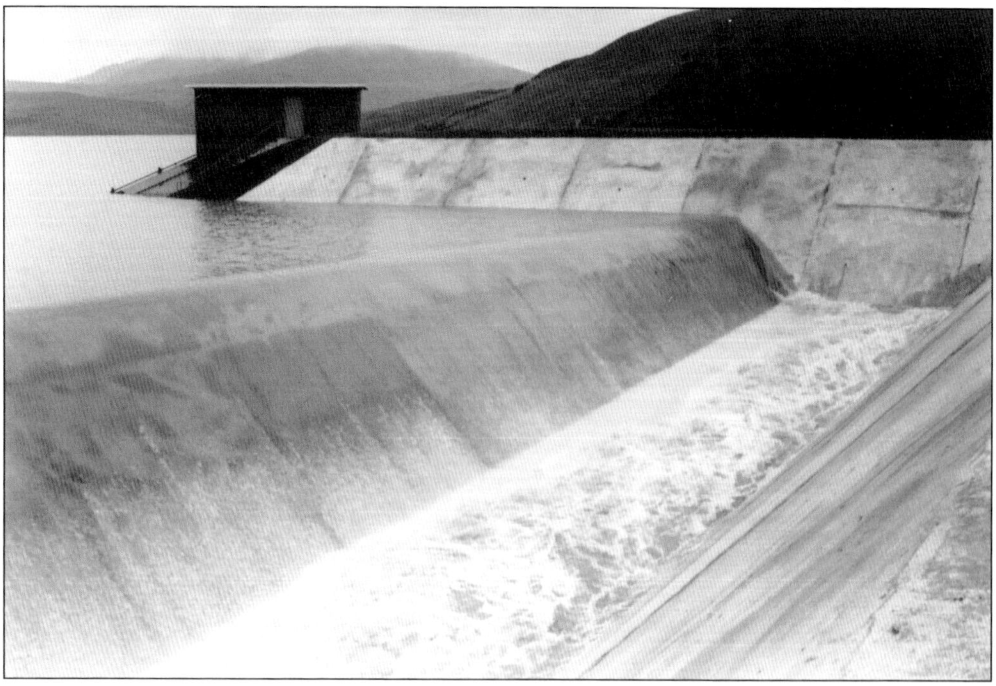

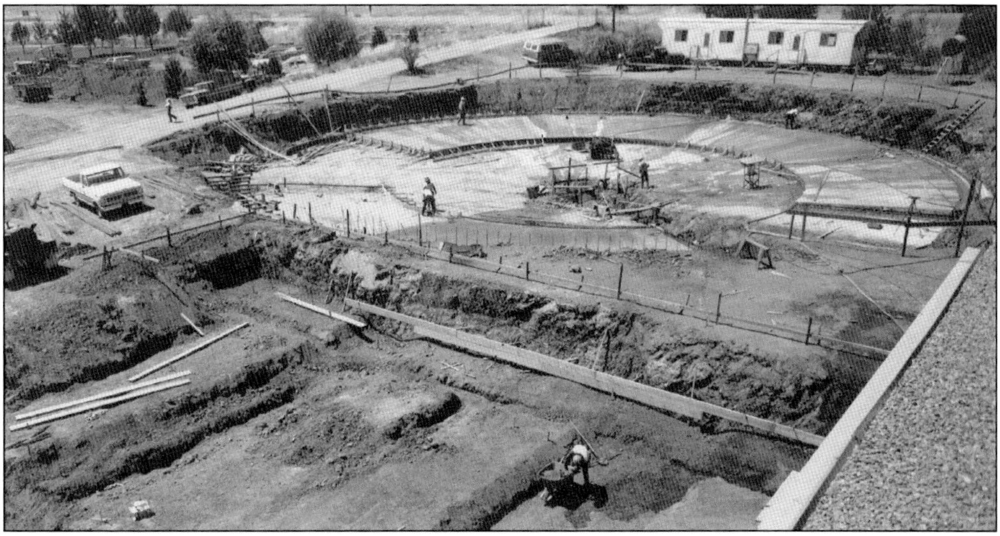

In the early days of MMWD, water was delivered directly from the mountain to the water users, but over time, water quality regulations became more rigorous. In the 1960s, water quality regulations required the district to filter the water. In response to those regulations, the Bon Tempe and San Geronimo Treatment Plants were constructed, and later expanded, keeping up to date with ever-improving water treatment regulations. These photographs show the sand filters and the flocculation and sedimentation tank at San Geronimo Treatment Plant in July 1972 (above) and September 1972 (below). Water treatment starts at the reservoirs, where the oxygen balance is maintained by aeration. At the treatment plant, suspended matter is removed by filters, and bacteria and pathogens are removed or inactivated by mechanical and chemical treatment. (Photographs by Construction Photographers.)

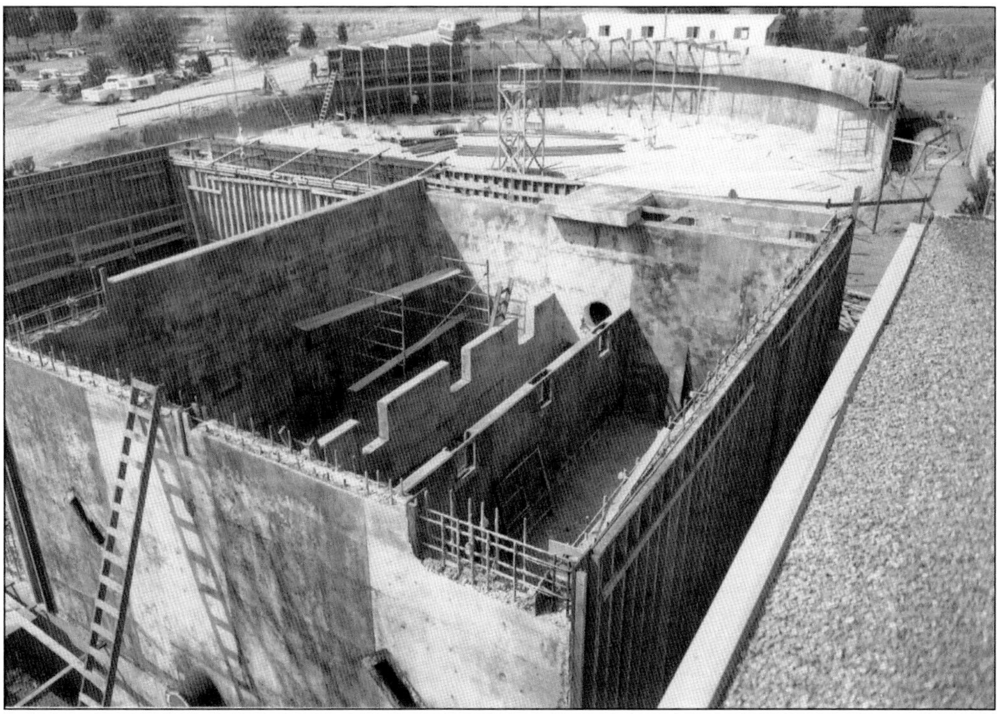

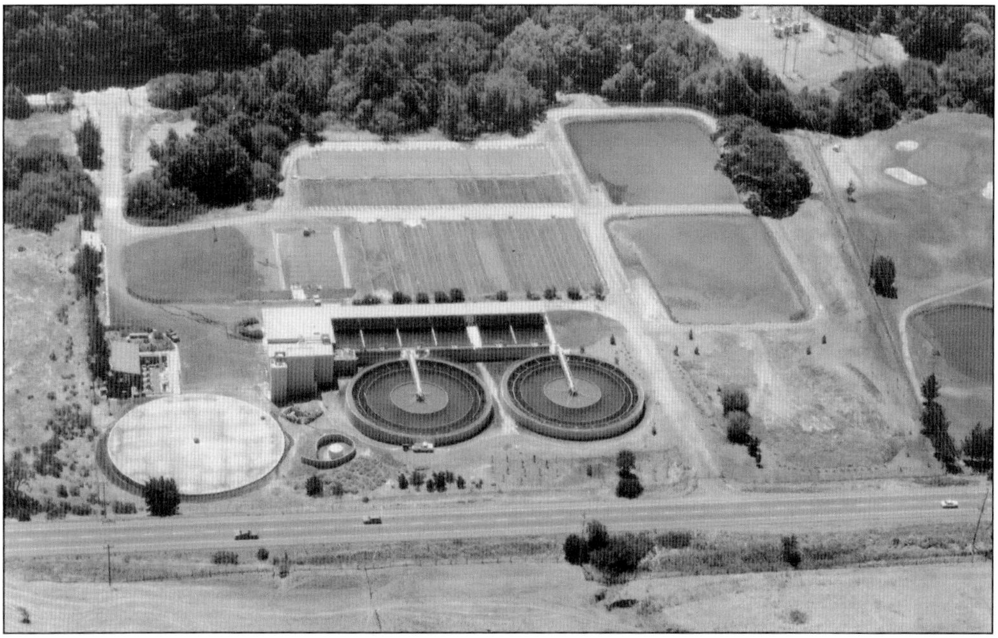

The San Geronimo Treatment Plant (shown here in 1974) and the Bon Tempe Treatment Plant, both for potable water, along with the Las Gallinas Treatment Plant for recycled water, are capable of processing 61 million gallons of water per day. (Photograph by Ed Brady, Aero Photographers.)

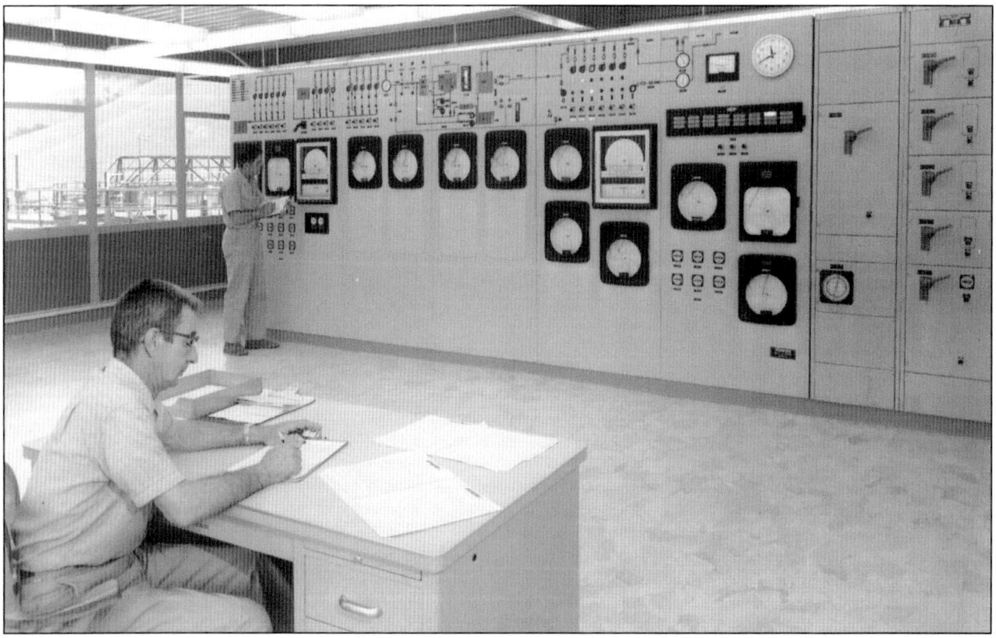

The MMWD system can deliver up to 41,000 gallons of potable water per minute. It typically takes about 24 hours for the water to travel from a reservoir to one of the treatment plants and then through many miles of underground pipe to water users. The reservoir levels, flows, and pump operations are monitored and controlled electronically. San Geronimo Treatment Plant's graphic control panel, which has since been updated with computerized controls, is pictured here. (Photograph by the Foxberry Company.)

Six

The District Outgrows Its San Rafael Roots

After the Marin Municipal Water District's charter in 1912, the board of directors met at the San Rafael Town Trustees' meeting room in the former county courthouse at Fourth and A Streets. Water system operations began in 1916, after the district acquired the delivery systems from private water companies along with several of the properties owned by those companies. In November 1916, the district took over the offices of the Marin Water & Power Company on Fifth Avenue in San Rafael between A and B Streets. The building had been built by the Marin County Water Company in 1878 as its headquarters and consisted of 1,500 square feet of office space, plus a meter shop, pipe shop, store room, stable, and garage. The space accommodated the MMWD staff of just eight people at the time. But by 1922, the district outgrew its small space and purchased property at 874 Fourth Street from the William Tell Coleman estate. The district constructed a new building and moved into the space in 1924. That building still stands today, but it is heavily masked by renovations.

Throughout the war years, the district's staff size remained relatively constant, with only slight growth. But as the district grew following World War II, more space was needed. In 1951, the district expanded the corporation yard to property located in Greenbrae (later incorporated into Corte Madera), which is still the current district corporation yard today. By the end of the decade, the district's San Rafael office was overcrowded. The human resource and engineering staff had taken over the board room; staff working on the Nicasio Reservoir project were working in a separate office on Lindaro Street, and part of the engineering staff had been relocated to offices across from the building on Fifth Avenue. In 1959, the board of directors authorized the development of a new headquarters building, and on April 30, 1962, the staff of approximately 85 employees (now about 235) moved to the current location in Corte Madera.

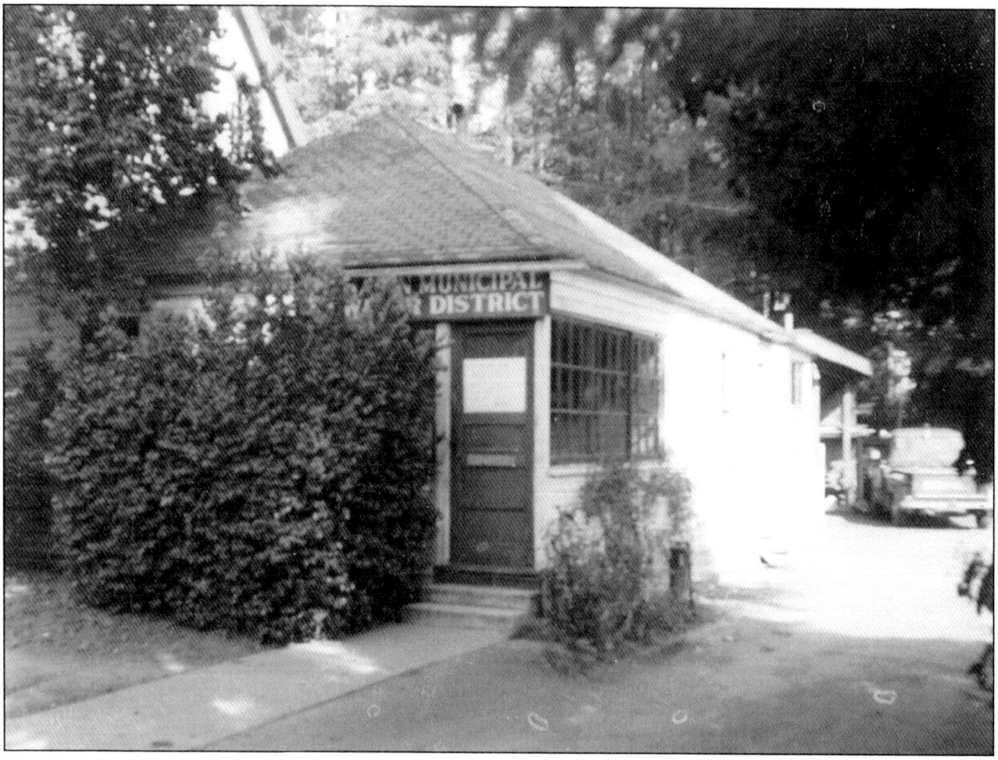

By 1916, four years after the creation of the Marin Municipal Water District, valuation of all the assets of the former private water companies was determined by the California Railroad Commission, the predecessor to the California Public Utilities Commission. With those determinations, MMWD bought most of the assets of those private companies. As a result, MMWD acquired an office and garage from the North Coast Water Company in Mill Valley on Corte Madera Avenue between city hall and what is now a grocery store. Built by the Tamalpais Land and Water Company in 1902, it was torn down for a city hall parking lot in 1959.

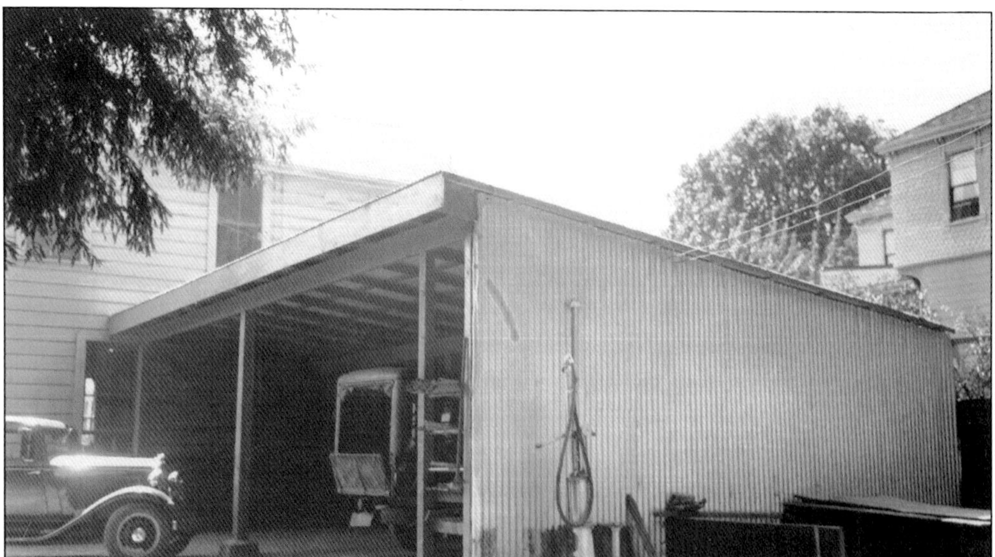

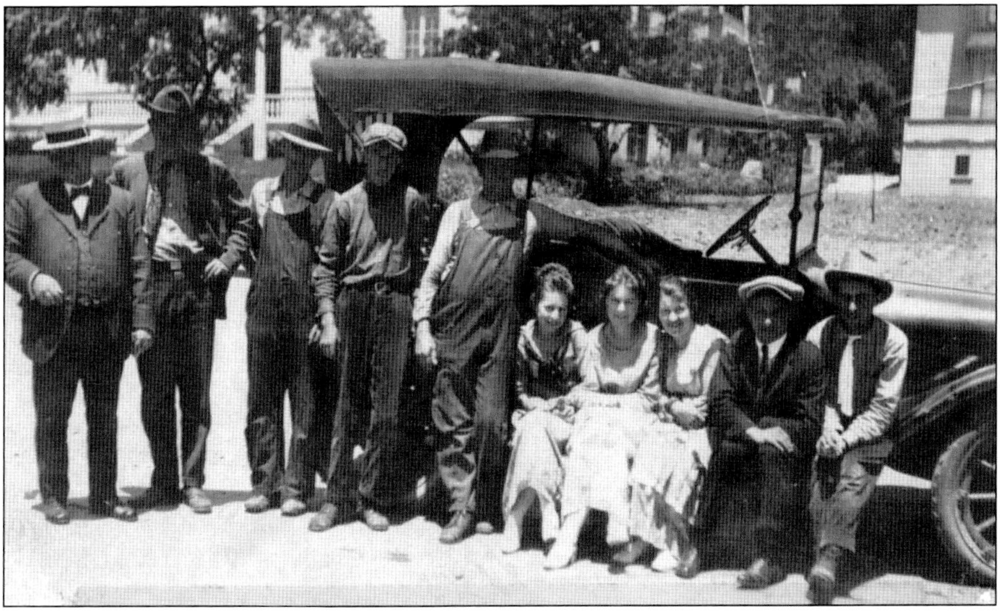

MMWD employees pose in front of the office at 425 Fifth Avenue in San Rafael around 1919. From left to right, they are Ed Prentice, Dick Longland, Joe Hughes, Joe Soldavini, Jack Mahar, Veronica Clark, Faye Clark, Ethel Longland, Louie Stephani, and Charlie Clement. The office was built in 1878 by Marin Water & Power Company. MMWD took over the office in November 1916 and occupied it until 1924.

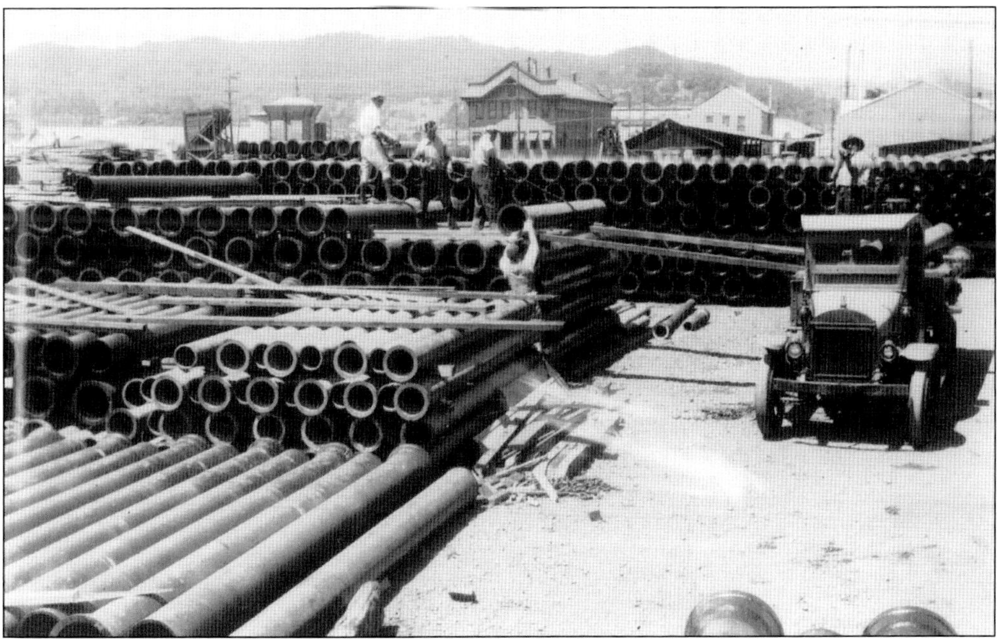

Men load pipe from the MMWD storage yard on Second Street and Grand Avenue in San Rafael in 1932. By that year, MMWD had outgrown that yard, which was purchased from the Marin Water & Power Company in 1916, and today the same functions are fulfilled from the site on Tamal Vista Boulevard in Corte Madera.

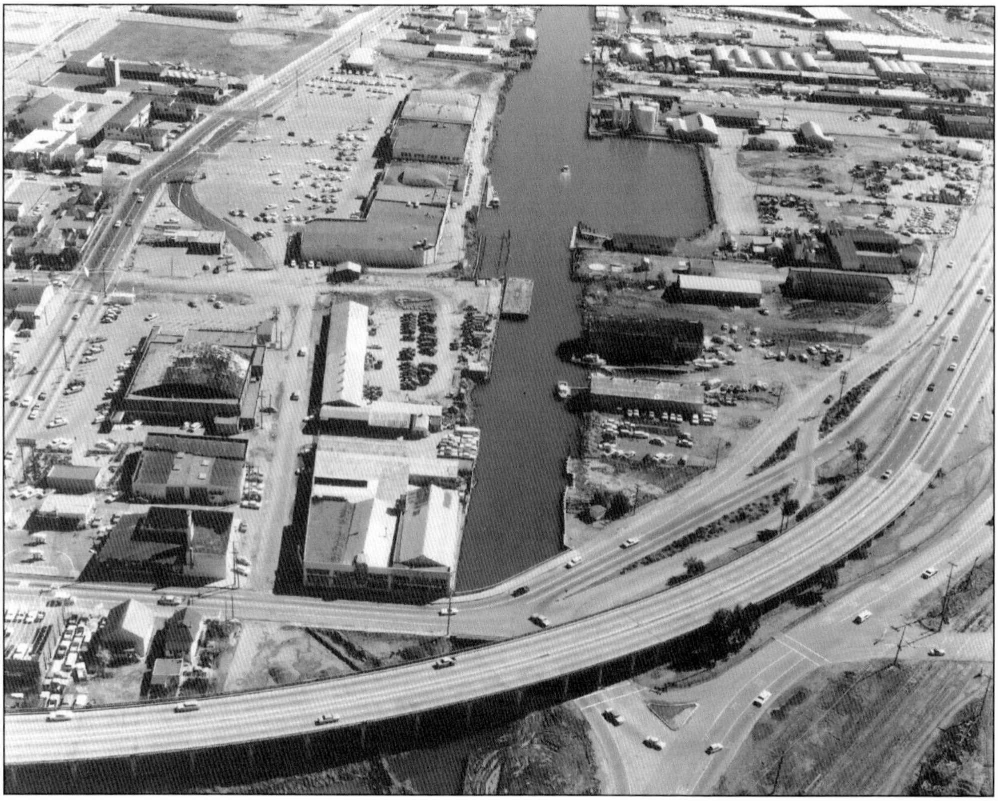

In 1927, the district purchased land for a new corporation yard from the Flood estate on the San Rafael canal at the southwest corner of Grand Avenue and Second Street. This above view, from 1965, shows the property (in the center of the picture) prior to construction of the bridge, which now crosses the canal at Grand Avenue. The below image shows the same corporation yard, which allowed for pipes to be barged up the canal and unloaded directly at the wharf. (Above photograph by Aero Photographers.)

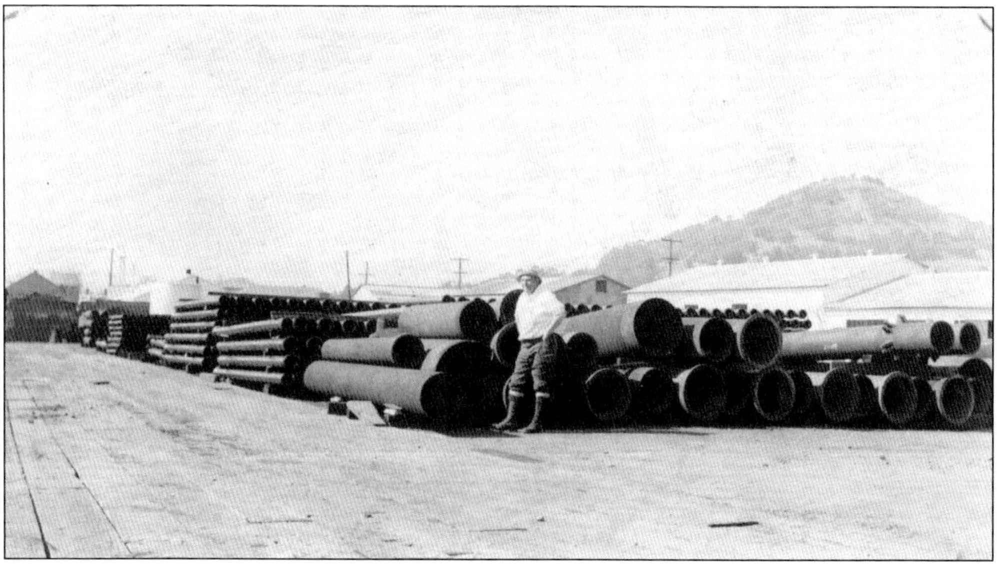

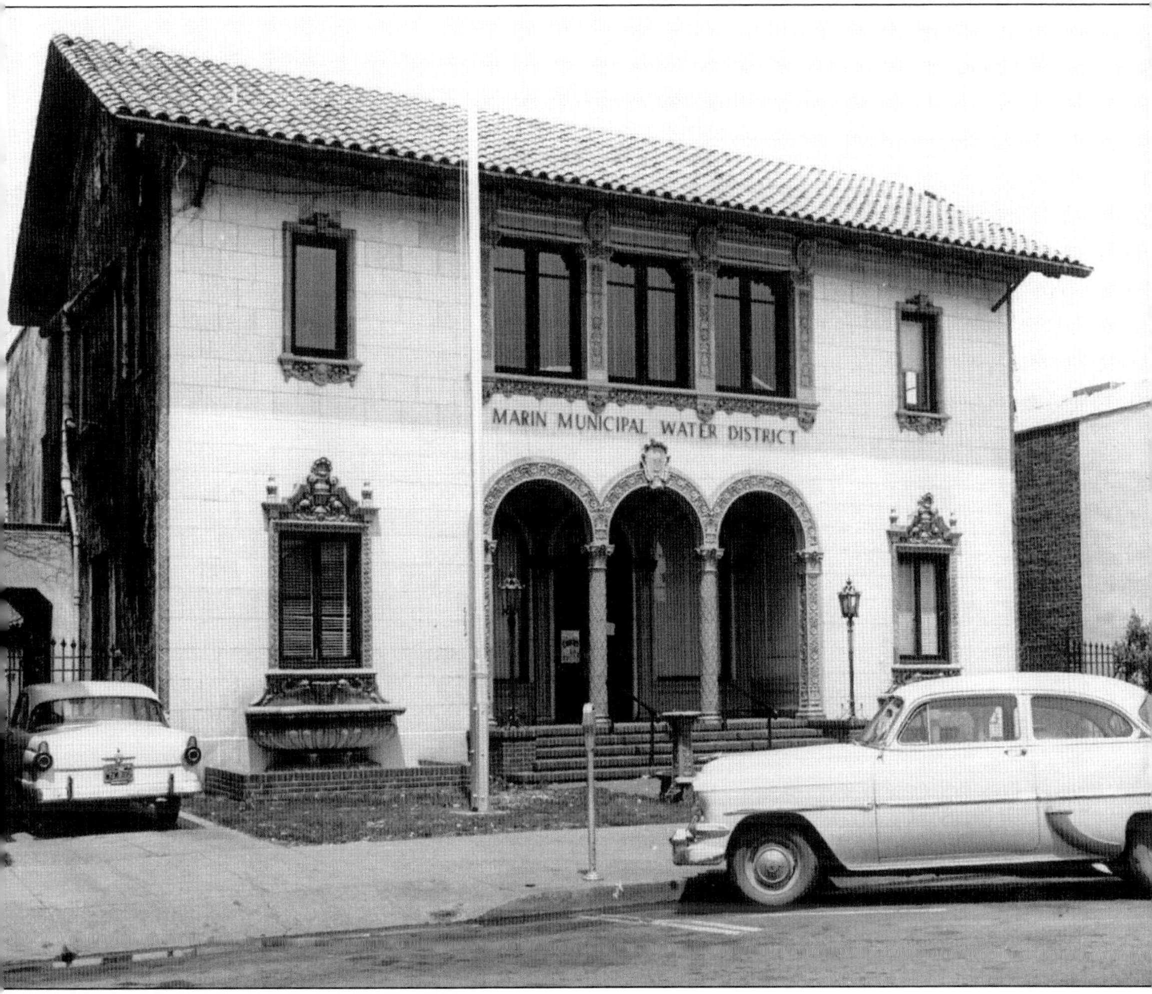

After the formation of the Marin Municipal Water District in 1912, water district meetings were held at the county courthouse in San Rafael at Fourth and A Streets. At that time, and for several years after, the district did not have its own offices. But by November 1, 1916, when it began operations, MMWD took over the former Marin Water & Power Company offices located on Fifth Avenue behind city hall and across from the Elks Club. MMWD soon outgrew the Fifth Avenue space, and in 1922 purchased the lot at 874 Fourth Street and contracted with Monson Brothers Construction to build a new office building. Today, the building is disguised by an addition on the front, and only the distinctive upper windows and roofline are visible from Fourth Street.

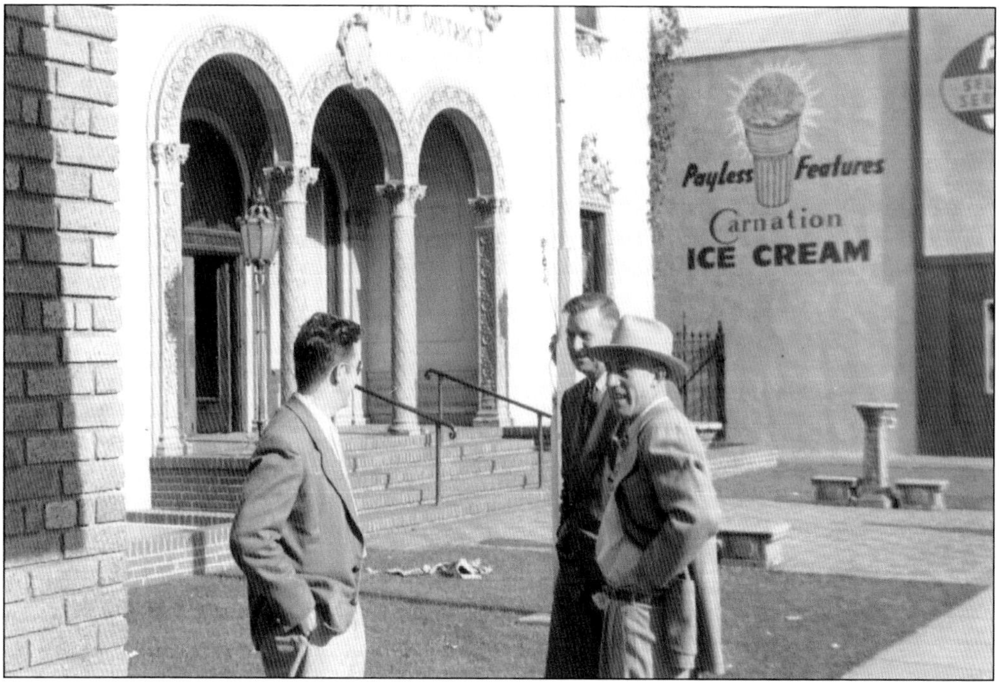

Pictured here in front of the former MMWD offices in San Rafael are, from left to right, Bill Brant, Stan Sacton, and Joe Soldavini. On Saturday, March 15, 1924, prior to moving into the new building at 874 Fourth Street, MMWD held a public reception as a showcase for the new facility. The following Monday was designated as "Moving Day," but by the end of the day on Sunday the staff had moved the offices from Fifth Avenue to the newly constructed building, and employees were ready for operations on Monday, March 17. Those offices, small by the standards of today, served as the district's offices from 1924 until 1962.

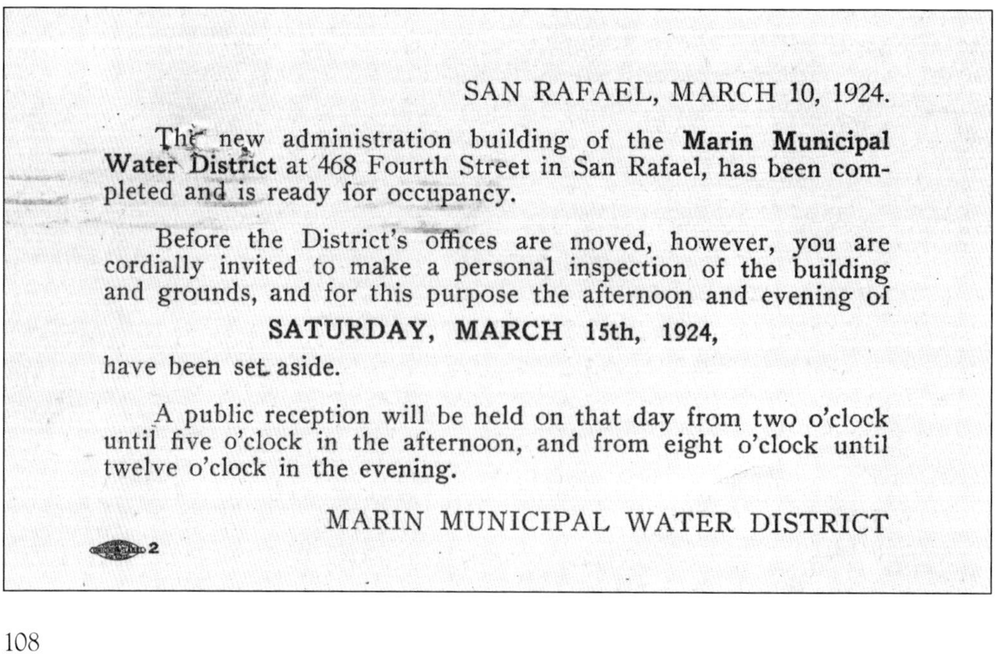

SAN RAFAEL, MARCH 10, 1924.

The new administration building of the **Marin Municipal Water District** at 468 Fourth Street in San Rafael, has been completed and is ready for occupancy.

Before the District's offices are moved, however, you are cordially invited to make a personal inspection of the building and grounds, and for this purpose the afternoon and evening of

SATURDAY, MARCH 15th, 1924,

have been set aside.

A public reception will be held on that day from two o'clock until five o'clock in the afternoon, and from eight o'clock until twelve o'clock in the evening.

MARIN MUNICIPAL WATER DISTRICT

2

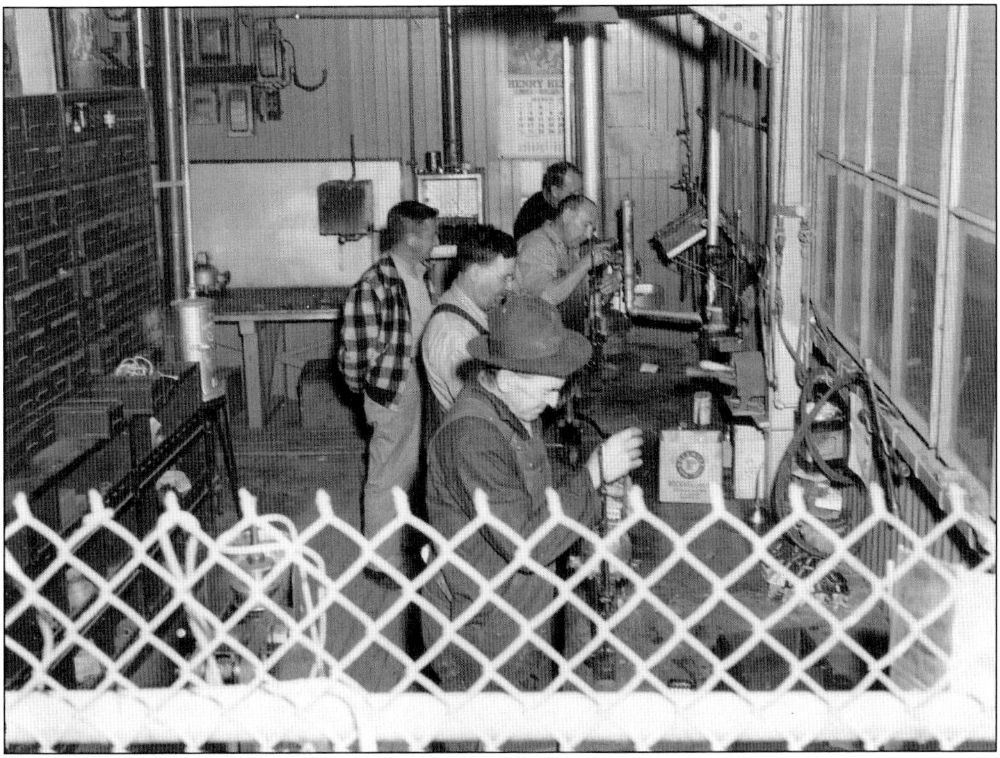

When it was built in 1924, the MMWD office building at 874 Fourth Street in San Rafael housed the general manager's office, other senior staff offices, the engineering department, a small meter shop (pictured above), supply room, billing room, lunch room, and a state-of-the-art customer service area (pictured right). It did not house a water quality laboratory, but after knowledge and awareness of water quality issues grew over the years, a lab was added. (Photograph at right by Bob Hax.)

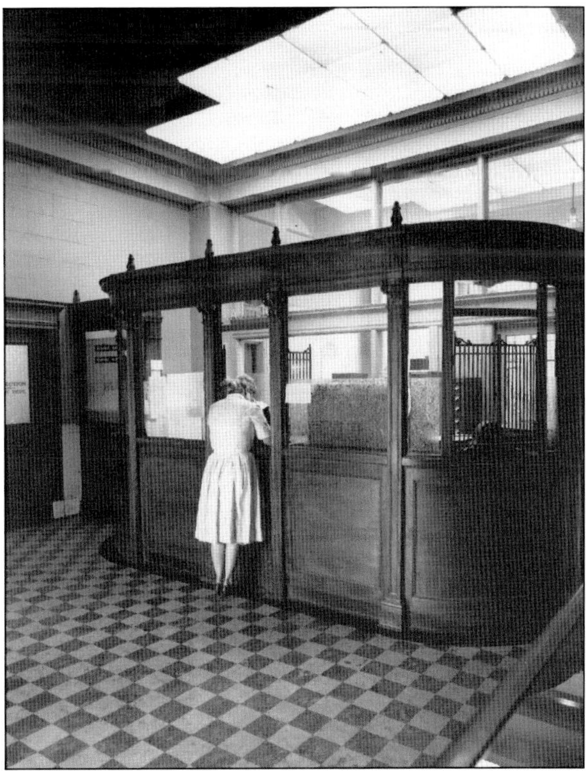

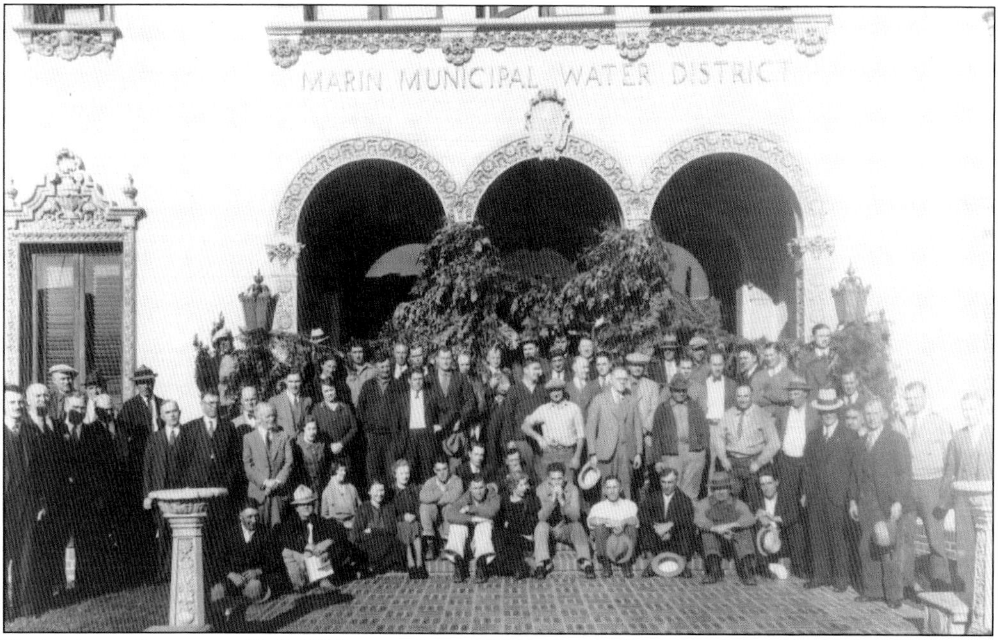

Both Marin County and MMWD grew during the 1950s and 1960s and water district employees were scattered in facilities across the county. One longtime employee remembered moving into the Fourth Street office in 1924 and wondering "who could have designed a building as large as that, as the furniture we had wouldn't begin to fill the rooms in the building." But space became less and less plentiful as MMWD grew, as seen in this photograph of employees posing outside the Fourth Street headquarters. In 1959, MMWD began a search for a new headquarters building, which included locations at Deer Park and Phoenix Lake picnic grounds and others at Terra Linda and the Marin Civic Center, but the current site in Corte Madera was selected, and employees packed and moved over the weekend (below) to be ready for business on Monday, April 30, 1962.

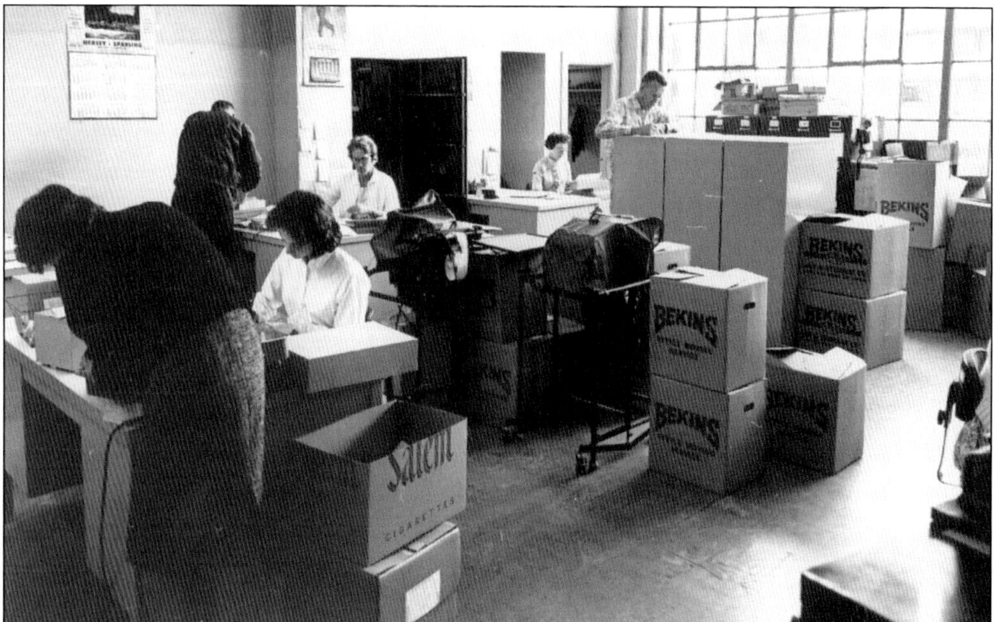

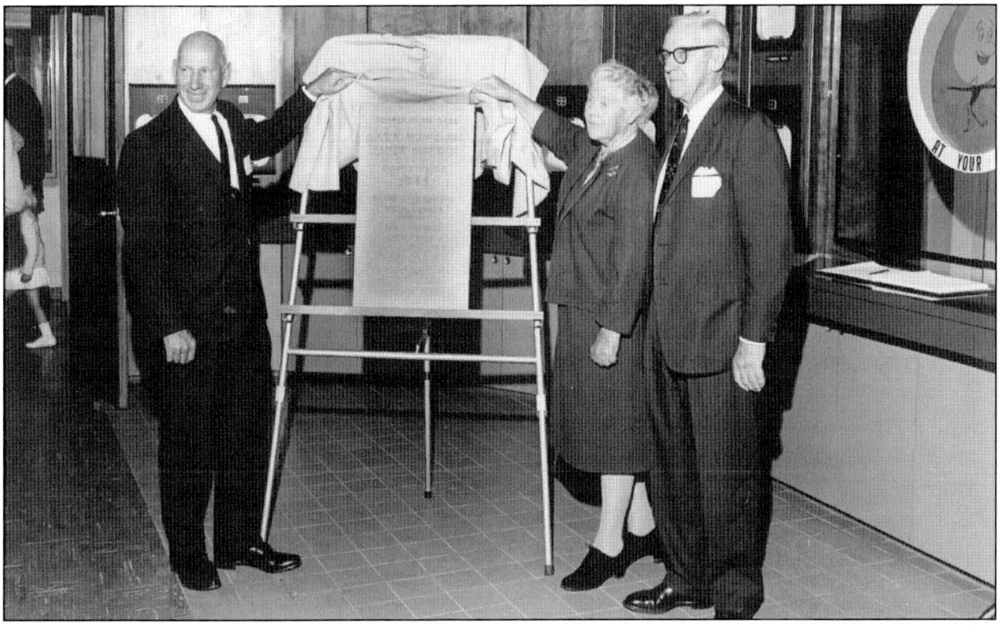

On August 29, 1956, the board of directors called for a $12.6 million bond measure to finance the Nicasio Dam, Bon Tempe and San Geronimo Treatment Plants, new water transmission facilities, and a new main office building for the water district. The measure was brought before the voters on November 6, 1956, and was overwhelmingly approved, with 81 percent voting in favor. After several years of exhaustive research, the board of directors approved the purchase of the current site at 220 Nellen Avenue in Corte Madera—then designated as the Greenbrae property. Two parcels were purchased from the Fifer family for $153,780. In March 1961, a $743,369 contract was awarded for construction of a 32,000-square-foot office building. On Sunday, May 13, 1962, a dedication ceremony was held.

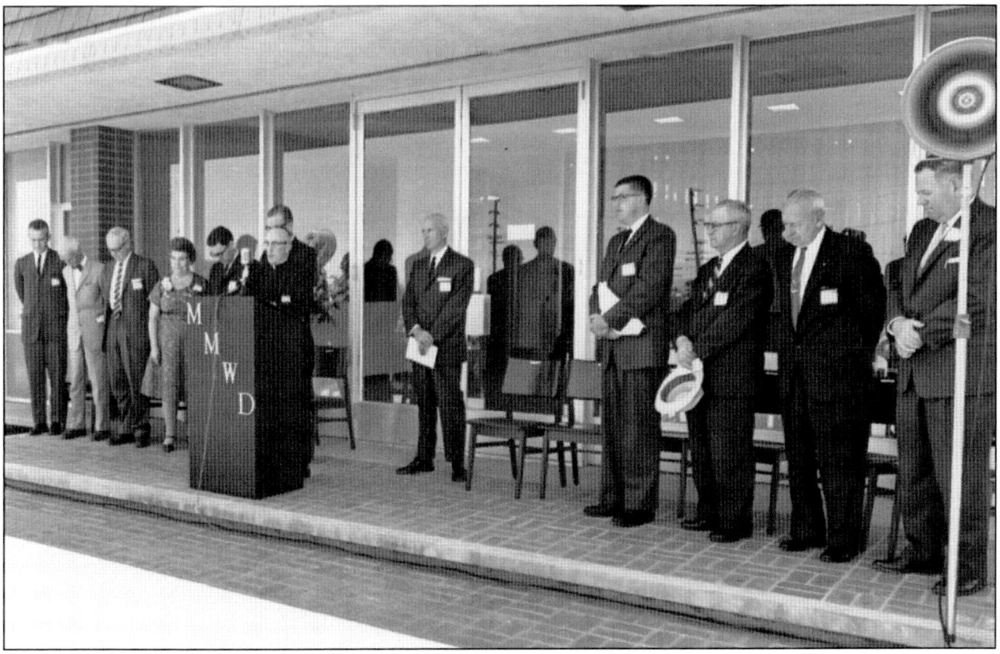

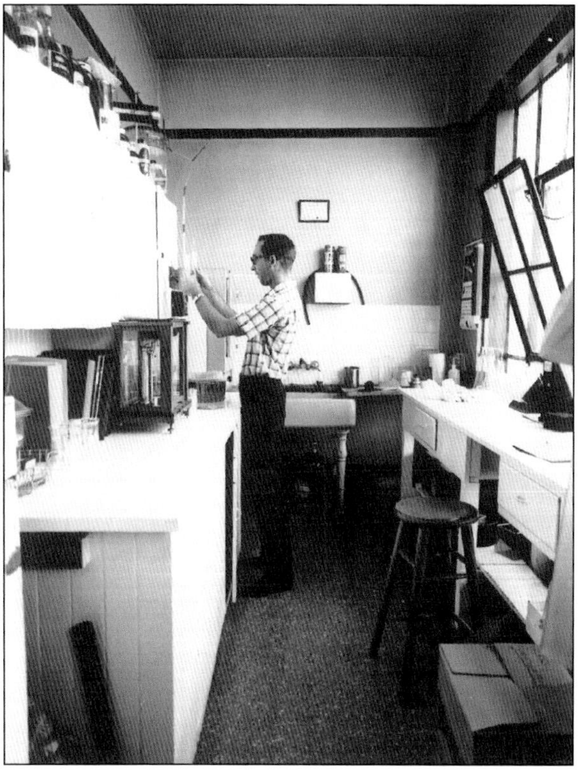

In 1924, when MMWD opened its offices on Fourth Street, no thought was given to installing a water quality laboratory. In that era, water quality depended on the source of the supply. The Lagunitas Creek source, remote from human habitation, was considered sufficient to satisfy water quality regulations of the day. But as standards changed, a water quality laboratory was installed (left). When the new Corte Madera building was dedicated, a water quality laboratory was part of the construction (below). In 2004, a new water quality laboratory, designed to meet the most modern standards, was built across the street from the main office. In a conscientious and complex process, MMWD's water quality laboratory conducts more than 120,000 water tests per year.

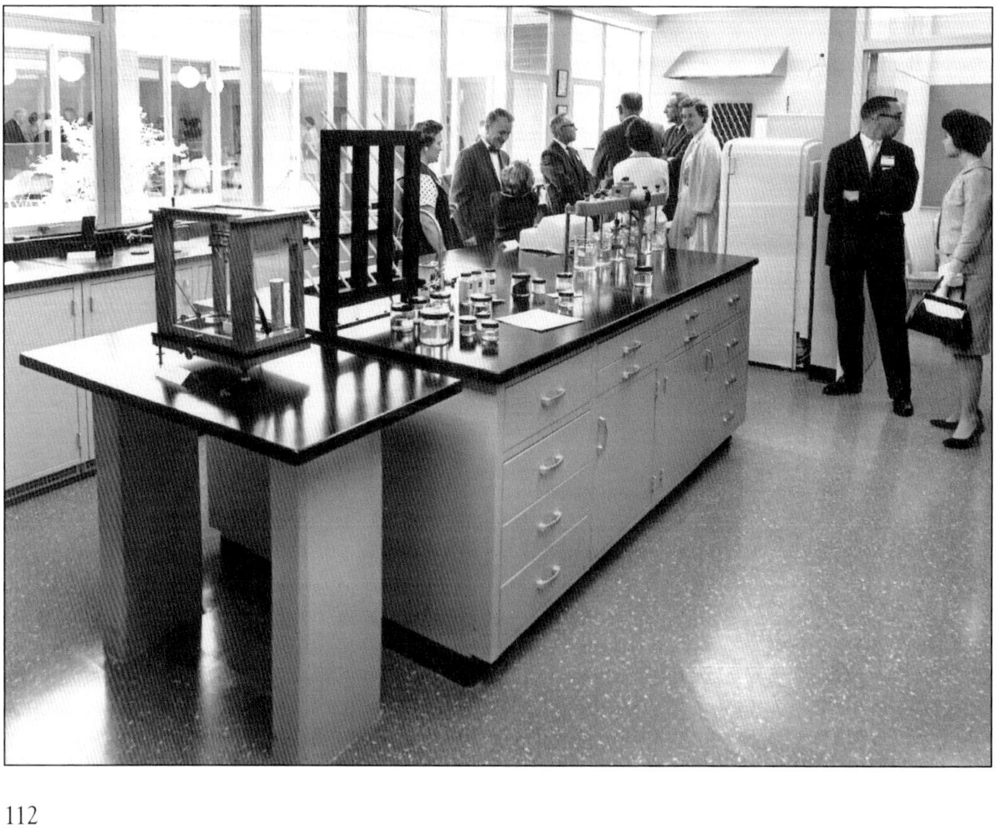

Seven

WATER SHORTAGES

The district's initial water supply report identified two potential water supply sources: the Lagunitas Creek watershed in Marin County and the Russian River in Sonoma County. Partly to keep water independence, the board of directors decided to develop the local supply from Lagunitas Creek. However, as pressure for additional water supply grew in 1970, the voters overwhelmingly approved the construction of the first phase of the Marin-Sonoma Aqueduct, or the "Intertie," designed to allow for water delivery via the Russian River. It would provide enough water to double the population of Marin.

The following year, the larger phase of the Russian River water supply project was brought to Marin voters, and funding was needed for Marin's share of Sonoma's Warm Springs Dam and the Marin-Sonoma Aqueduct. But the political climate had changed, and the voters were more concerned with growth and development than future water supply shortage. Using slogans such as, "More Water Means More People" and "Don't Los Angelesize Marin," the voters defeated the proposal for Russian River water by almost 90 percent.

A few years later, in 1976 and 1977, Marin suffered the worst drought in its history. The district's reservoirs were essentially empty, drawn down to mere pools well below operational levels. The district implemented mandatory water rationing with water use reduced to 37 percent of pre-drought levels and prohibited new water hookups. Additionally, an emergency plan was implemented that allowed Marin to tap into the State Water Project with a temporary pipeline placed across the Richmond-San Rafael Bridge.

In 1978, the district brought forward a proposal to build one last Marin dam at Soulajule in West Marin. Against the background of the drought, the voters approved the proposal by a four-to-one margin. The district attempted to keep the water supply and demand in balance after the drought by relying on extensive water conservation, the use of water recycling, and an increase in Russian River water imported through the Intertie, as negotiated under the 1970 voter approval.

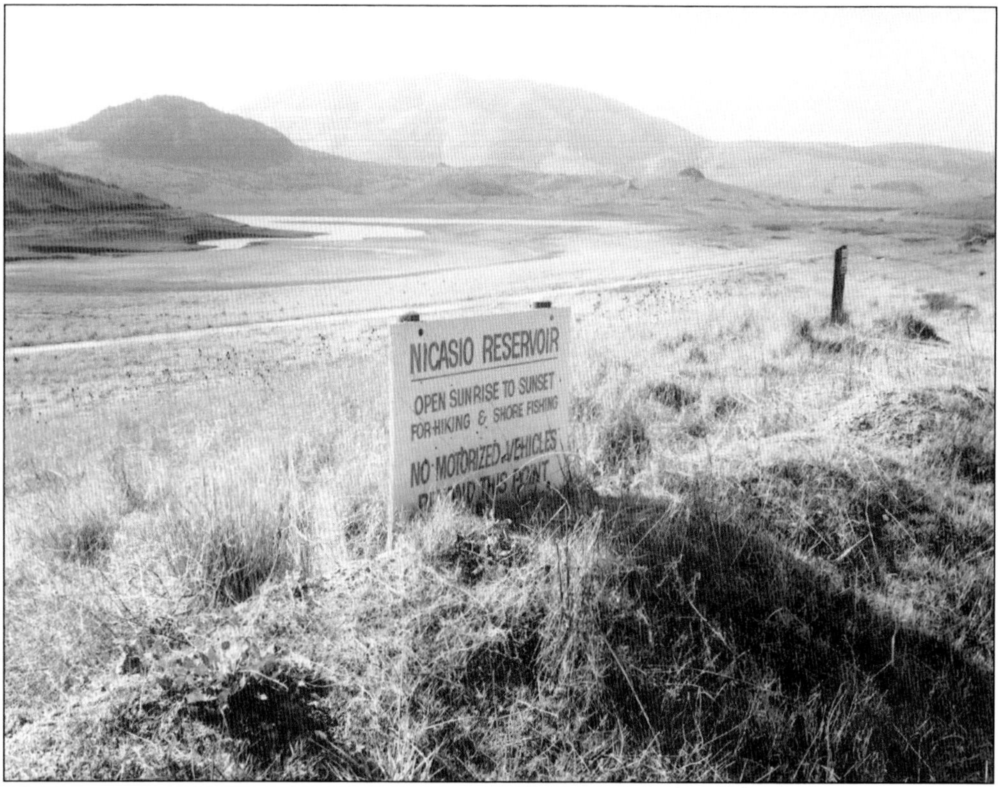

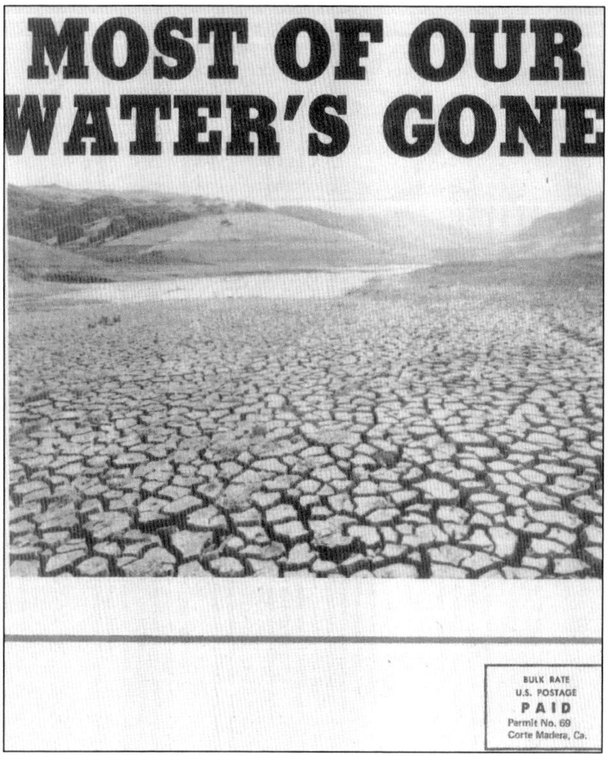

In the mid-1970s, MMWD was still entirely dependent on the rainfall captured in local reservoirs for water supply. Rainfall during the winter of 1975–1976 was only 55 percent of normal, and the following season it was only 44 percent of normal. In response, the board of directors passed a mandatory rationing plan in early 1976. As the drought continued into the second year, a more stringent plan, requiring users to cut their water use by 57 percent, was implemented. Even with such tight restrictions, only 25 percent of the water supply remained and the rationing plan would only have stretched reservoir supply to December. These two photographs of Nicasio Reservoir, one from a district mailer to customers, dramatically demonstrate the severity of the situation. (Above photograph by Richard Wheeler.)

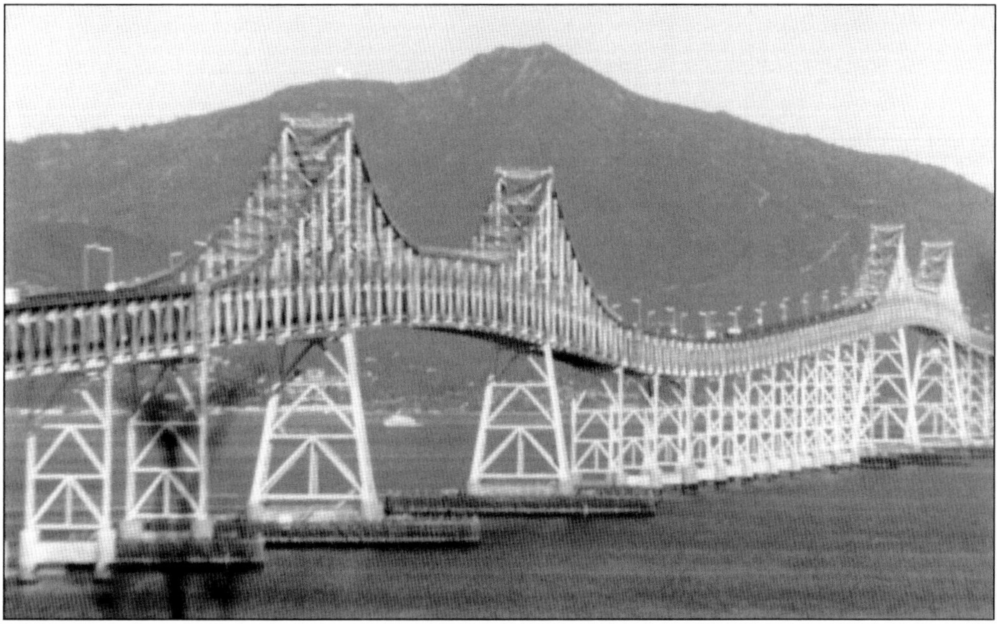

Realizing that the county was at risk of being out of water very soon, MMWD reached out to other water districts and the California Department of Water Resources and in February 1977, the Metropolitan Water District of Southern California agreed to reduce its use of water from the State Water Project so that water could be used to address droughts elsewhere in the state. MMWD was a principal beneficiary of that agreement, and an emergency temporary pipeline was built on the 5.5-mile-long Richmond-San Rafael Bridge to bring the water to Marin. Workers in the below image unload segments of the 24-inch-diameter pipe that was used.

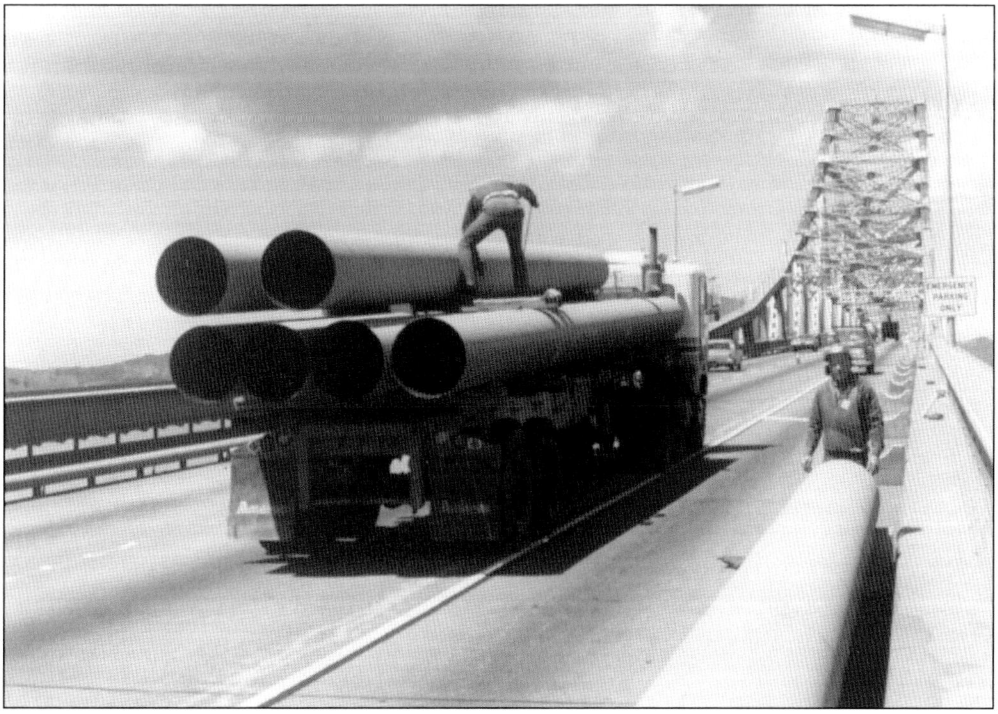

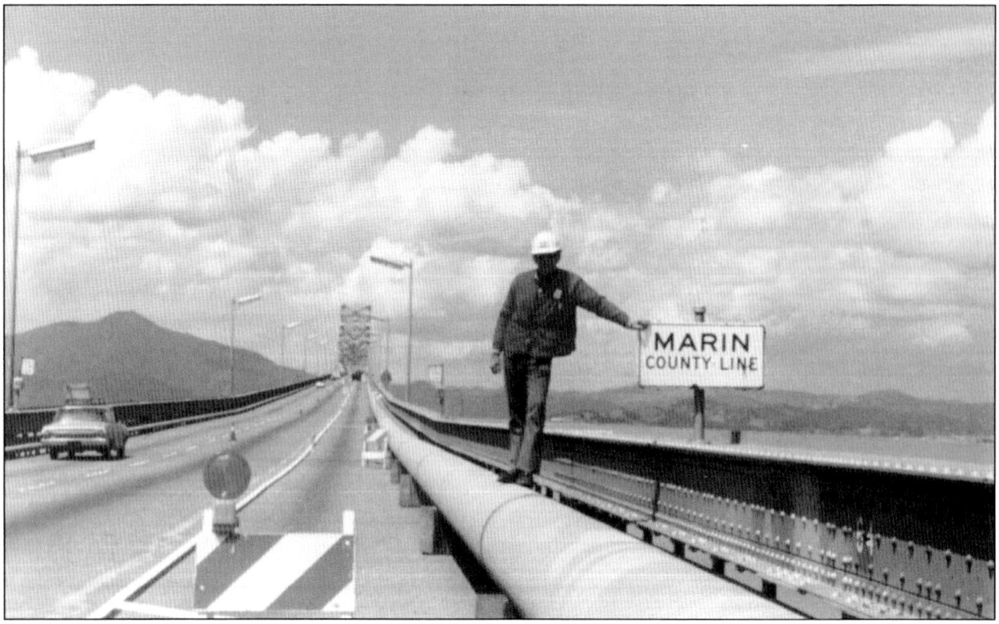

MMWD engineering inspector Jim Vitek poses on the emergency pipeline that was built across the Richmond-San Rafael Bridge in 1977. Between the additional water from the pipeline and the remaining reservoir water, MMWD water users were able to get through the remainder of the drought, conserving even more than the 57 percent mandatory rationing amount required by MMWD.

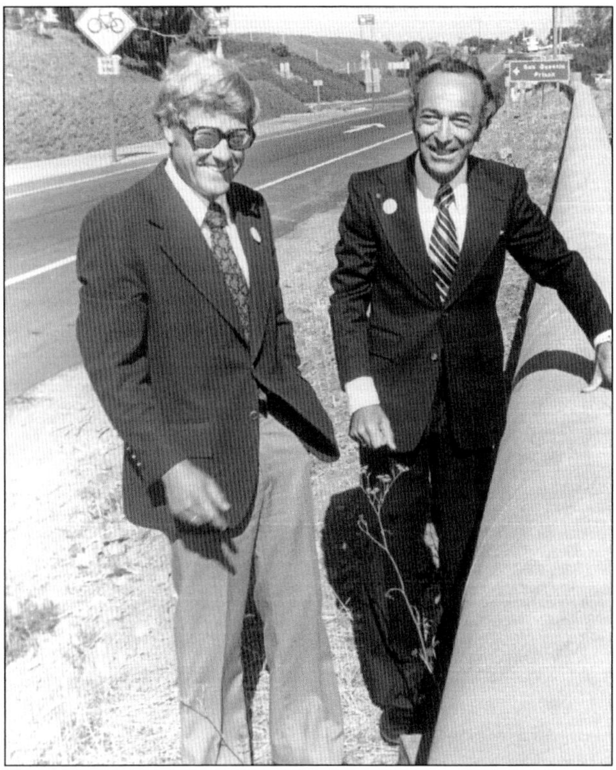

MMWD general manager Dietrich Stroeh (left) and director Bill Filante stand next to the completed emergency pipeline on the Marin County side of the Richmond-San Rafael Bridge. The first water through the pipe arrived in San Rafael on June 7, 1977. MMWD contracted for 10,800 acre feet—or 3.5 billion gallons—of water but used just 4,595 acre feet (1.5 billion gallons) between June and December 1977, when drought-breaking rain fell. (Courtesy of Dietrich Stroeh, photograph by Edmond Scola.)

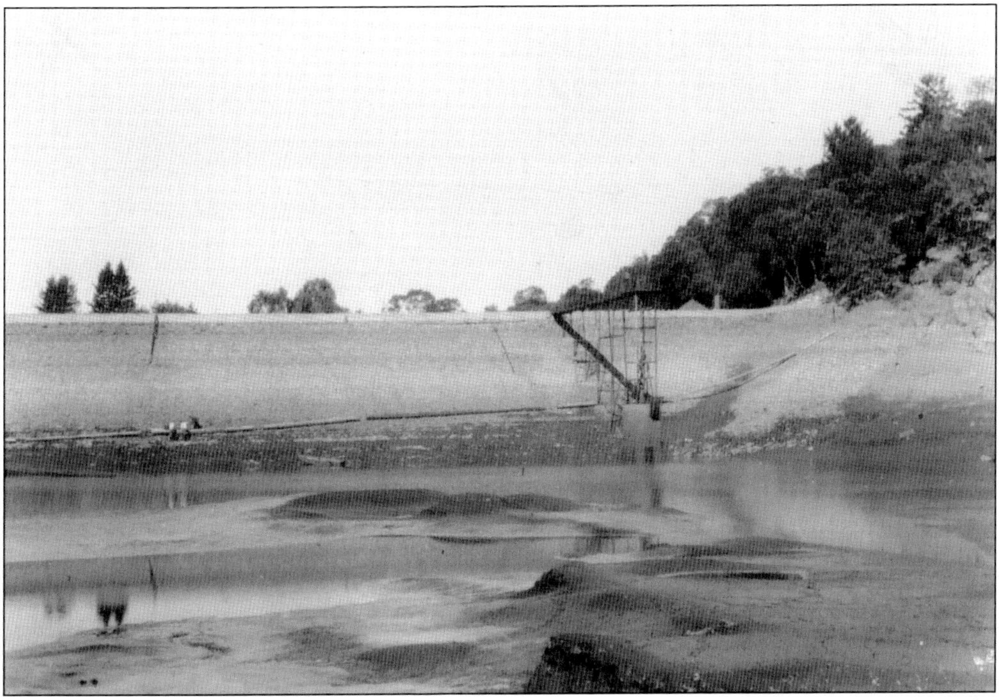

The season of 1976–1977 is considered the drought of record in Marin. However, during the early-to-mid-1930s, Marin experienced a severe drought that lasted six years and emptied out Lagunitas Lake in this 1936 photograph. The 1930s drought was mitigated by the fact that MMWD managed supply for a much smaller population.

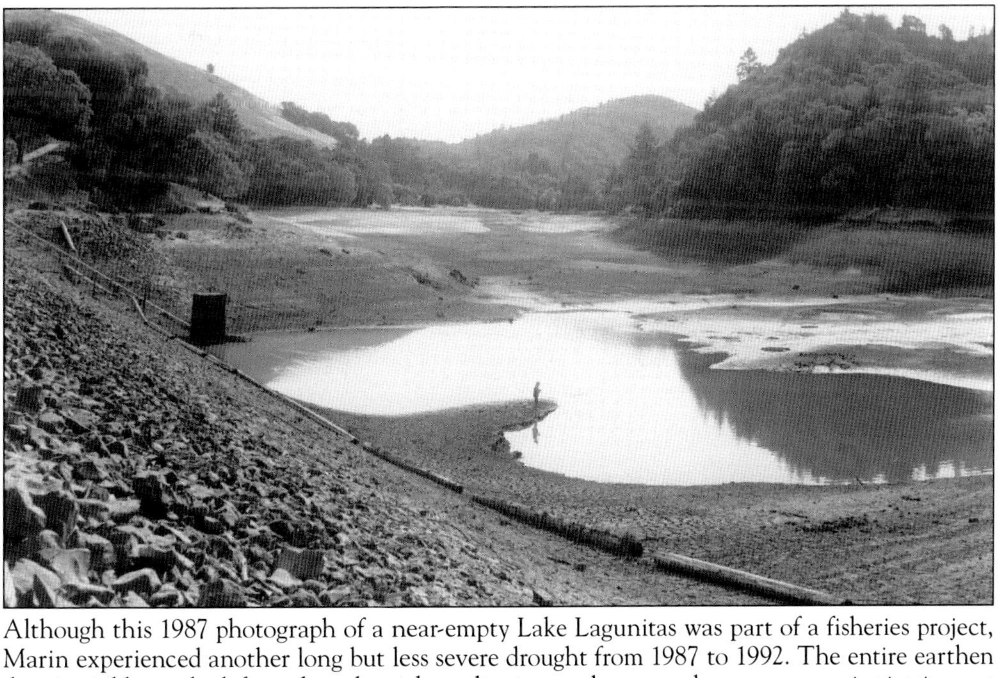

Although this 1987 photograph of a near-empty Lake Lagunitas was part of a fisheries project, Marin experienced another long but less severe drought from 1987 to 1992. The entire earthen dam is visible on the left, and on the right, a district employee can be seen on a point jutting out into the water. The reservoir was refilled a few months after this photograph was taken.

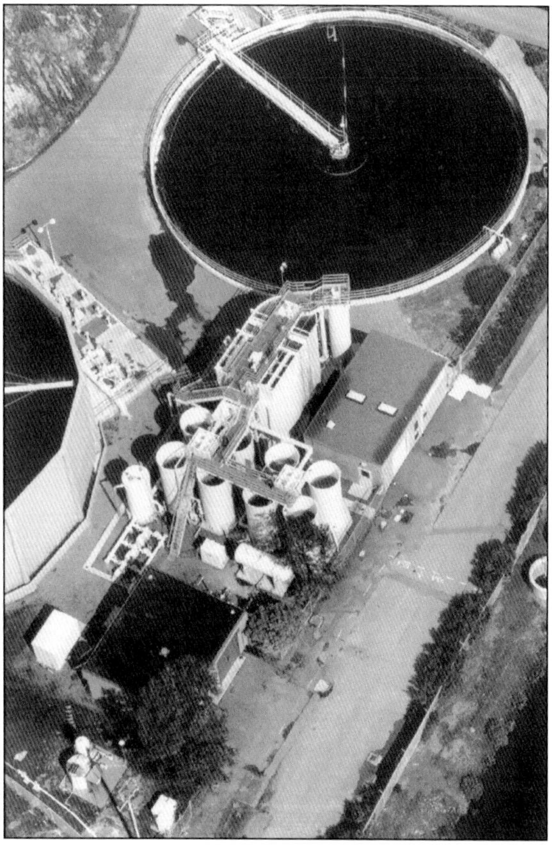

MMWD built a pilot recycled water treatment plant in 1975 that was used to provide irrigation water during the 1976–1977 drought. Then, in 1981, MMWD built a permanent plant to supply irrigation water and reduce the demand for potable water. In 1990, MMWD doubled the capacity of that plant (pictured here) to two million gallons per day and upgraded the treatment process to produce the highest quality recycled water possible. Working with Las Gallinas Valley Sanitary District, MMWD has provided recycled water throughout the area of the sanitary district and is working on plans to expand the recycled water effort southward to the Peacock Gap area of Marin. (Photographs by Richard Wheeler.)

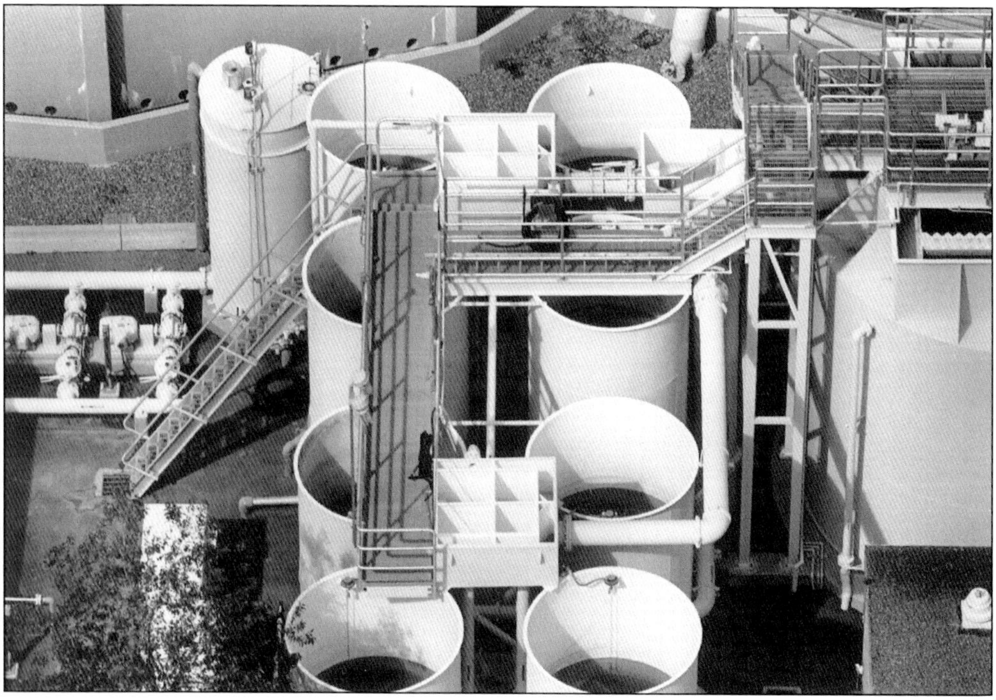

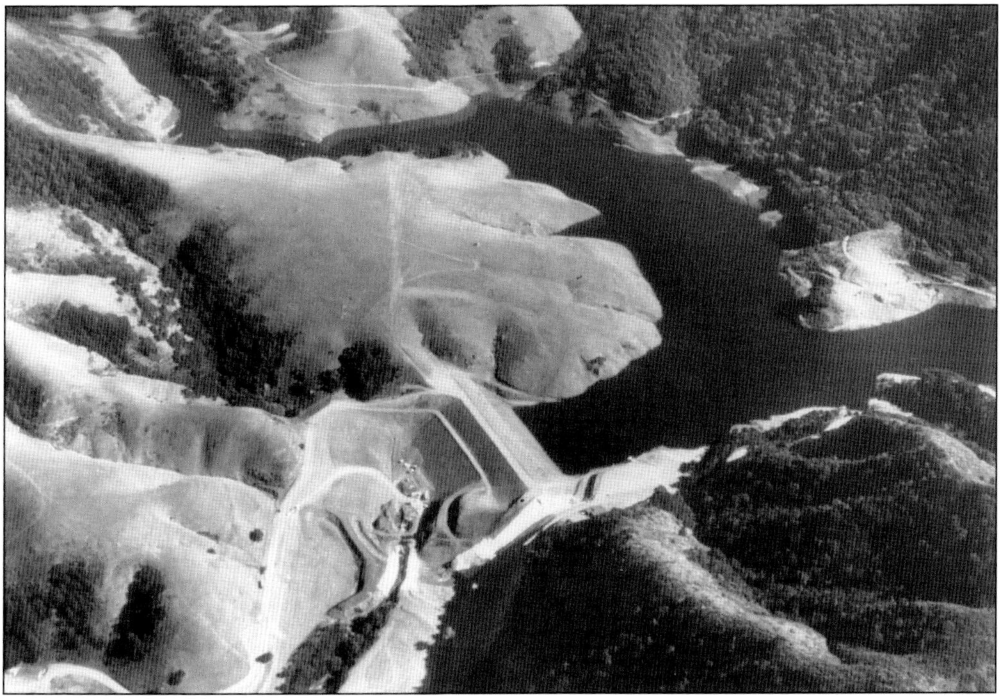

Located even farther west than Nicasio Reservoir in the rolling hills of West Marin, MMWD built Soulajule Reservoir in 1979 as a reserve supply to be used in a drought. The reservoir was used as a water supply source during the 1987–1992 drought. It is used only during droughts because the water has to be pumped the farthest of any of the reservoirs to reach a treatment plant, so it is more expensive to use. (Above photograph by Aero Photographers; below by Richard Wheeler.)

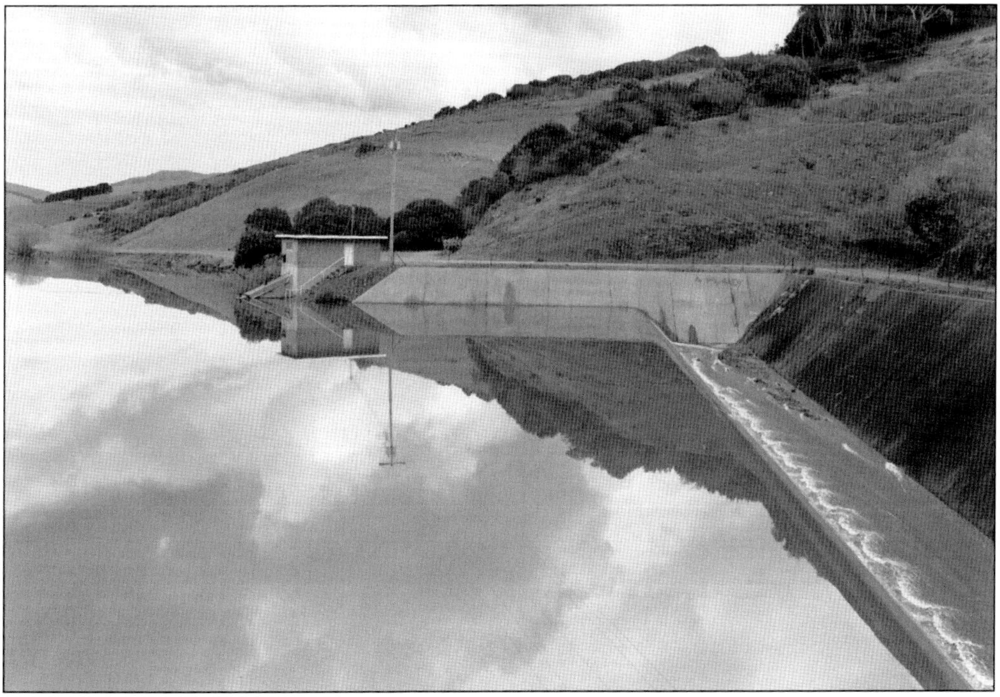

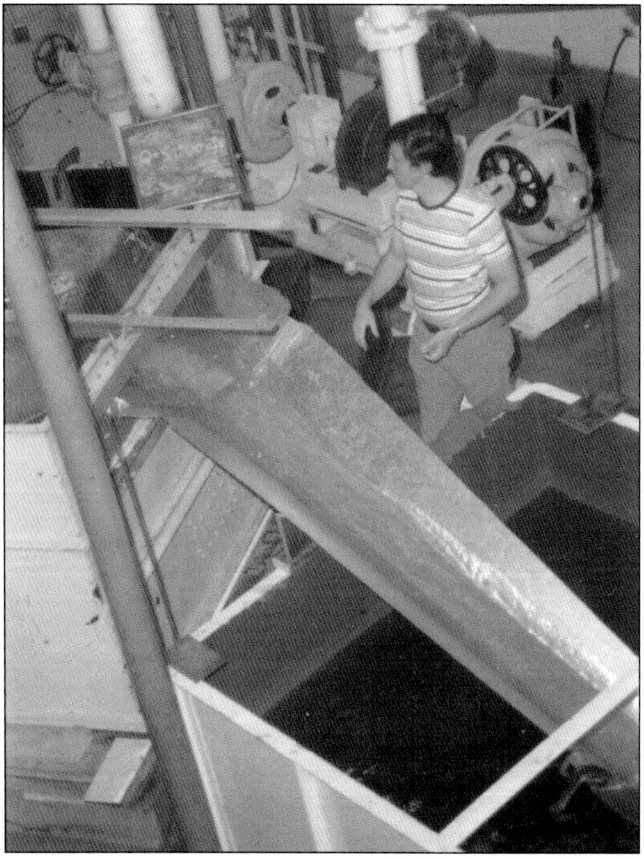

At left, a model of the Soulajule Dam spillway was made and tested at the University of California, Berkeley, as part of the design and construction process for building the reservoir. Below is a photograph from 1979 of the actual Soulajule spillway with water gushing down it. The water is flowing into a tributary of Walker Creek, which then flows west into Tomales Bay and the Pacific. During normal operations, water only makes it over the spillway when the reservoir is full and overflowing, but MMWD can open valves to release water into the spillway for testing or other purposes.

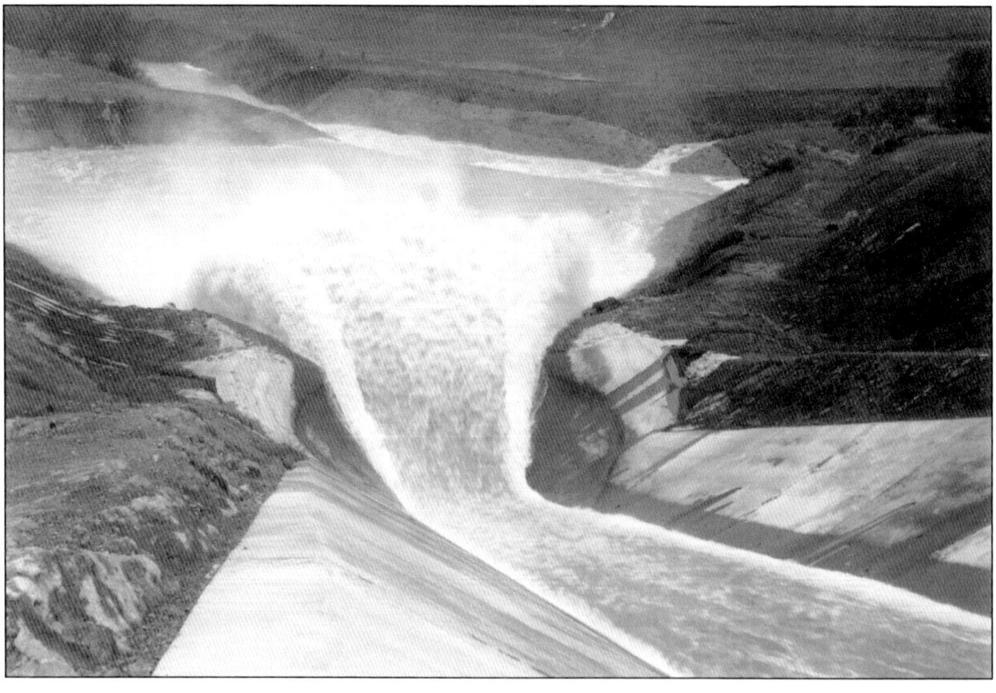

Eight

The District's Second Century

Mount Tamalpais and its watershed are part of the Golden Gate Biosphere Reserve. With more than 900 species of plants and 400 species of animals, many of which are rare, threatened, or endangered, the Mount Tamalpais watershed is recognized internationally as a biodiversity hotspot.

The Marin Municipal Water District has always recognized that its responsibility goes beyond providing the community water supply. The district continues to manage the natural resources provided by Mount Tamalpais and recognizes that a clean and healthy water supply can only come from a balanced and well-maintained ecological environment.

In 1995, the district developed the Mount Tamalpais Vegetation Management Plan to provide a policy guide to minimize wildfires, to control invasive species, to restore native habitats, and to preserve beneficial ecosystem functions on the watershed. Under the plan's direction, MMWD staff constructs and maintains fuel breaks, removes weeds, conducts controlled burns with local firefighters, and monitors significant resources. In 2008, the district began the process of updating this plan, with an eye toward the challenges posed by a changing climate.

Downstream of the district's reservoirs on Mount Tamalpais, Lagunitas Creek provides one of the most important spawning and rearing grounds for coho salmon and steelhead trout in California—both of which are on the federal endangered species list. In 1997, the district developed the Lagunitas Creek Sediment and Riparian Management Plan, a 10-year endeavor to enhance the habitat for much of the watershed's threatened wildlife. In 2011, continuing that effort, the district developed a new Lagunitas Creek Stewardship Plan.

Taking advantage of the great appreciation that the Marin community has for the Mount Tamalpais watershed, the district works with hundreds of volunteers annually to restore habitat, maintain trails, and monitor vegetation and wildlife.

The district maintains the reservoirs as the heart of the community water supply. In order to protect the water source and balance the supply against excessive demand, the district continues to develop extensive conservation programs. Using a water conservation code, a recycled water program, educational workshops, and other innovative conservation techniques, MMWD has managed to reduce the community water use by 32 percent since 1987.

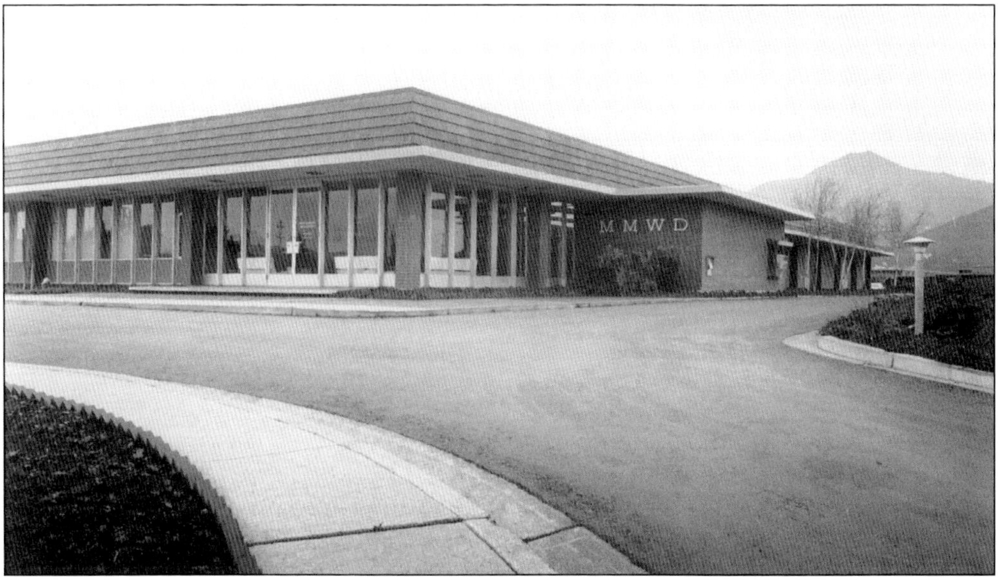

In 1962, after outgrowing its office building on Fourth Street in downtown San Rafael, MMWD built and moved to its new office at 220 Nellen Avenue in Corte Madera. The new administration building was constructed facing the freeway, US Highway 101, for easy visibility and access. MMWD has been in this location ever since, making building updates as needed to keep up with changing business practices. MMWD also built a new corporation yard, including a warehouse and office space, in 1962. This facility is located on Tamal Vista Boulevard directly across from the back of the Nellen Avenue administration building. In 2004, a new state-of-the-art water quality laboratory was built in front of the corporation yard (below). The property has an exceptional view of Mount Tamalpais. (Photograph by Barbara Stenson.)

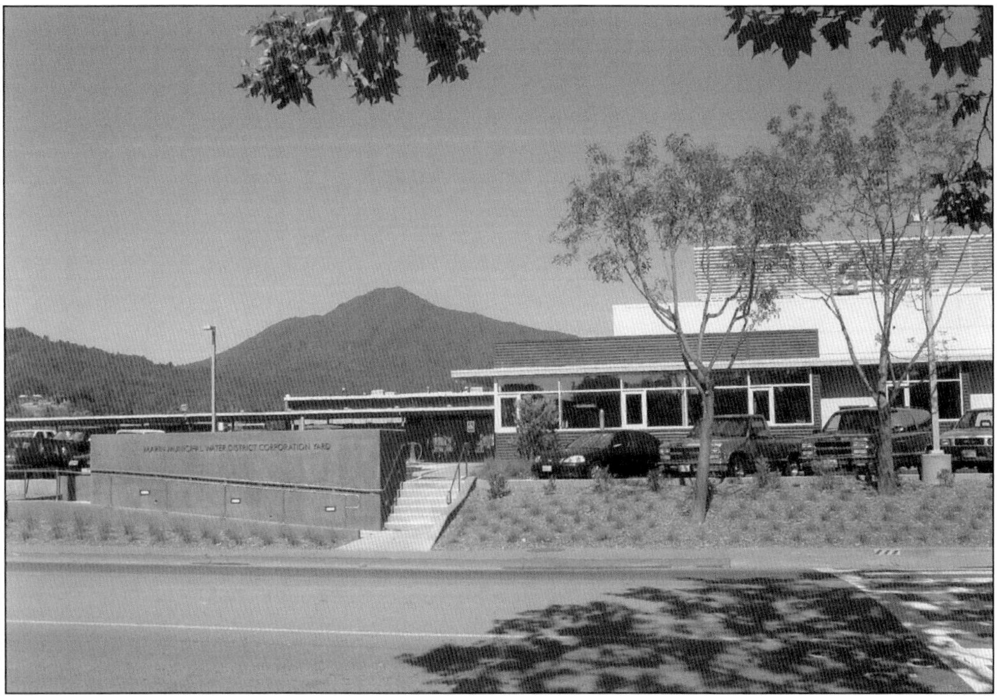

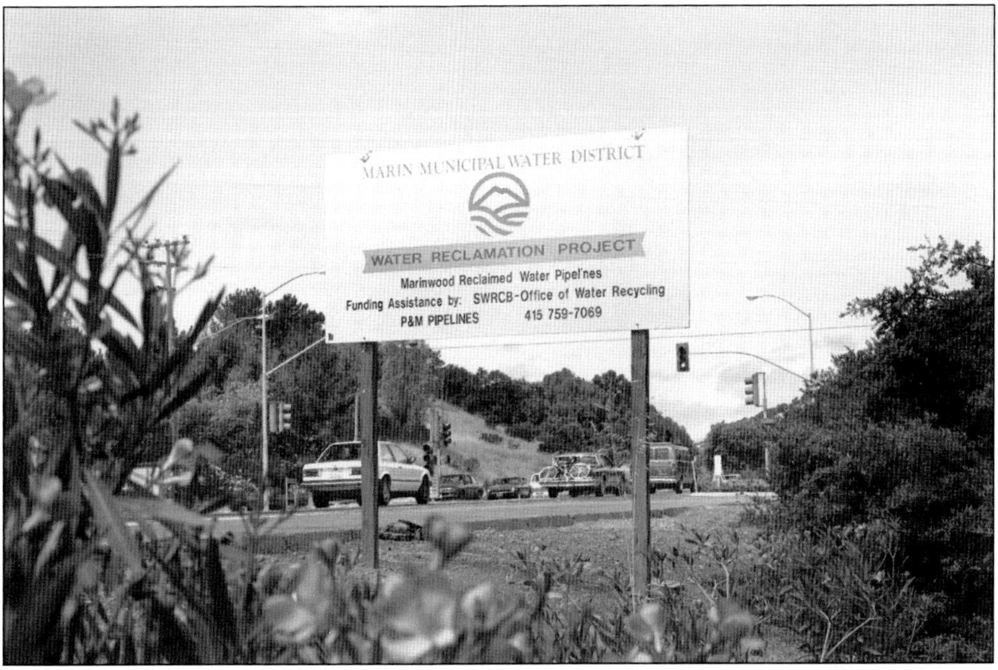

MMWD has an extensive recycled water system that serves more than 340 recycled water customers with 25 miles of pipeline, three water storage tanks, five pump stations, and one recycled water treatment plant. The recycled water program annually reduces the demand for potable water by the equivalent of the water usage of 1,800 single-family homes. Recycled water is used mostly for landscape irrigation. (Photograph by Richard Wheeler.)

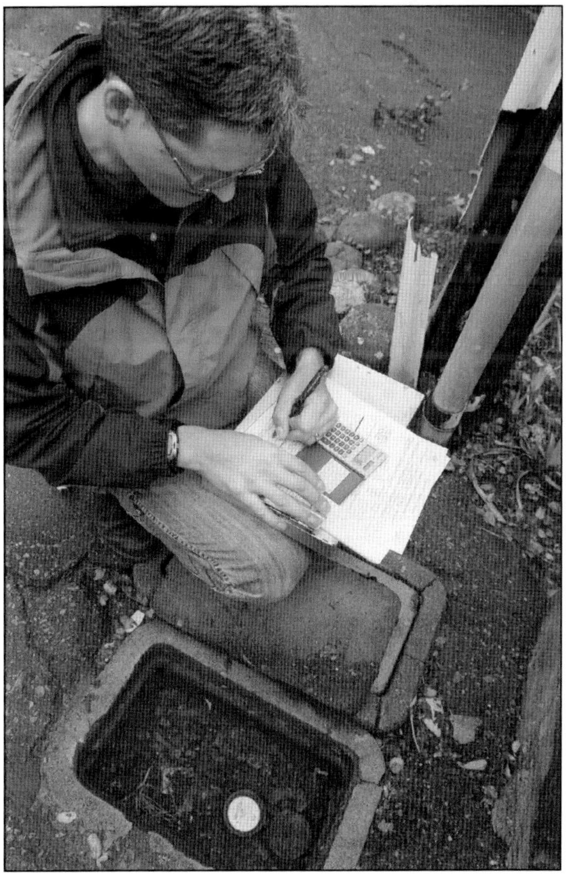

MMWD has an active water conservation program that includes consultations at customers' homes or businesses to help reduce water use. Every consultation includes a check of the customer's water meter, as MMWD Water Conservation Specialist Greg Van Trump does here, to check for water leaks. If no water is being used by the customer, and the meter indicates water movement through the pipe, a leak is likely. (Photograph by Richard Wheeler.)

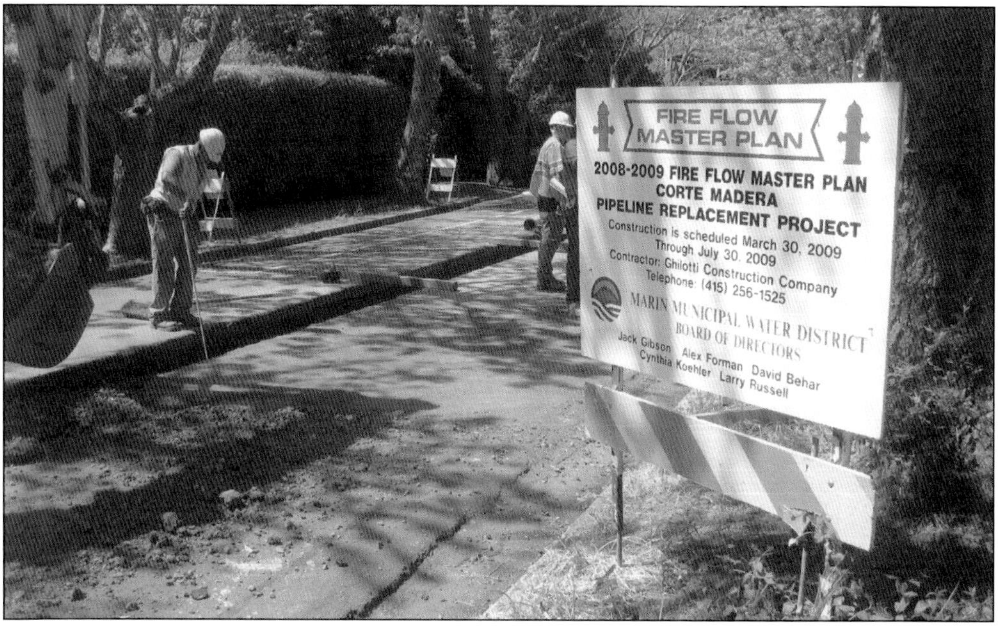

In 1997, MMWD embarked on the 15-year Fire Flow Master Plan, the district's first large-scale, community-wide plan to make improvements to the water system specifically to enhance the system's fire fighting capacity. Since 1997, MMWD has replaced 60 miles of undersized pipe and installed 750 hydrants in various locations. The plan also included seismic retrofitting of key facilities.

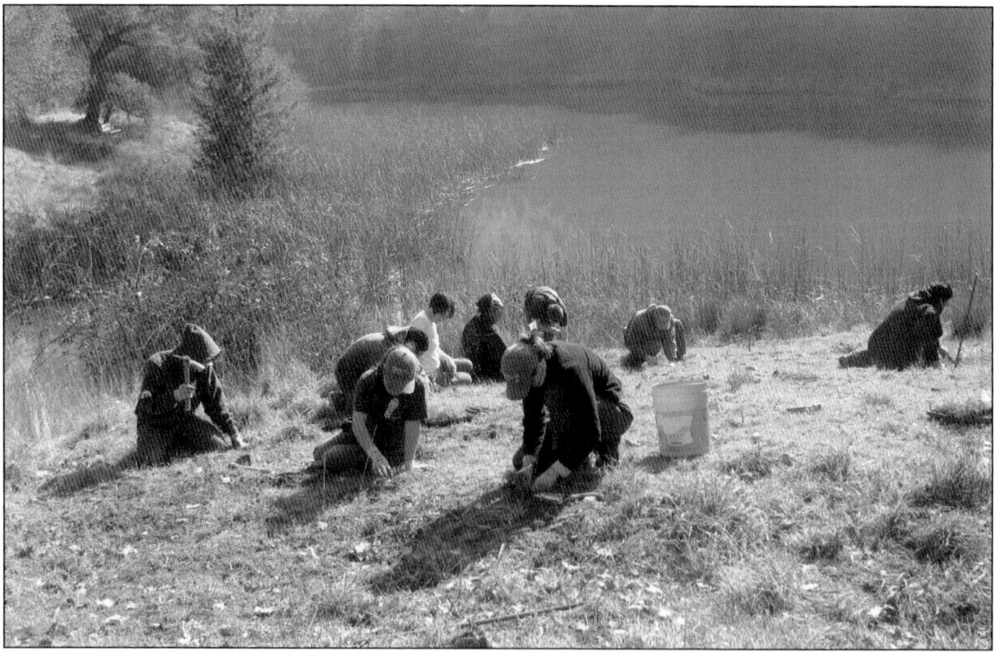

MMWD established its volunteer program in 1995. Volunteers play a critical role in maintaining the scenic beauty and biological richness of the Mount Tamalpais watershed. Most volunteer hours are devoted to restoring habitat, as Conservation Corps North Bay members are doing here, but other volunteer activities include trail restoration, vegetation monitoring, wildlife monitoring, mapping, fishery surveys, data analysis, and visitor education.

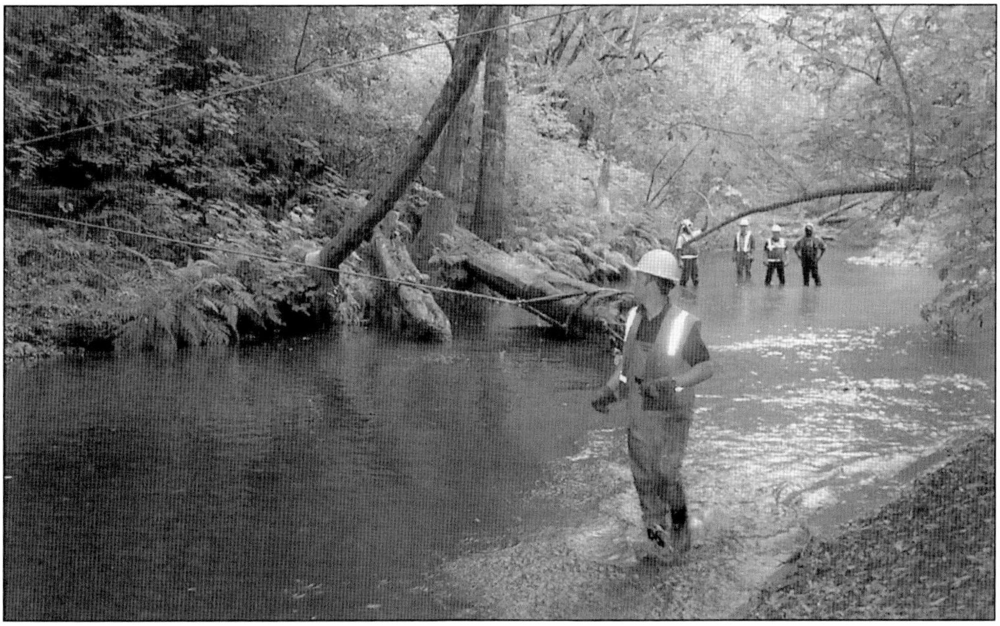

MMWD manages the habitat of Lagunitas Creek for the benefit of coho salmon, steelhead trout, and California freshwater shrimp with the goals of increasing and stabilizing the populations of all three species. Restoration activities include monitoring the numbers of coho and steelhead that migrate from the creek out to the ocean and maintaining flows and improving water quality for adult fish to swim upstream to spawn.

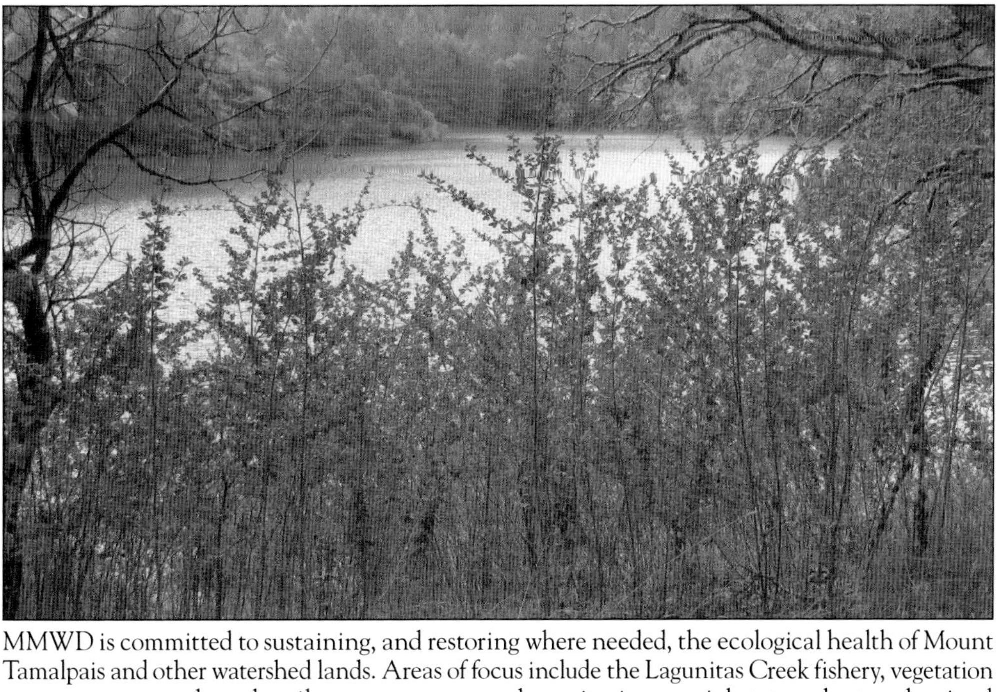

MMWD is committed to sustaining, and restoring where needed, the ecological health of Mount Tamalpais and other watershed lands. Areas of focus include the Lagunitas Creek fishery, vegetation management, roads and trails management, and monitoring special-status plant and animal species. The above image shows the proliferation of French broom, an invasive, non-native plant that increases wildfire risk and degrades native ecosystems.

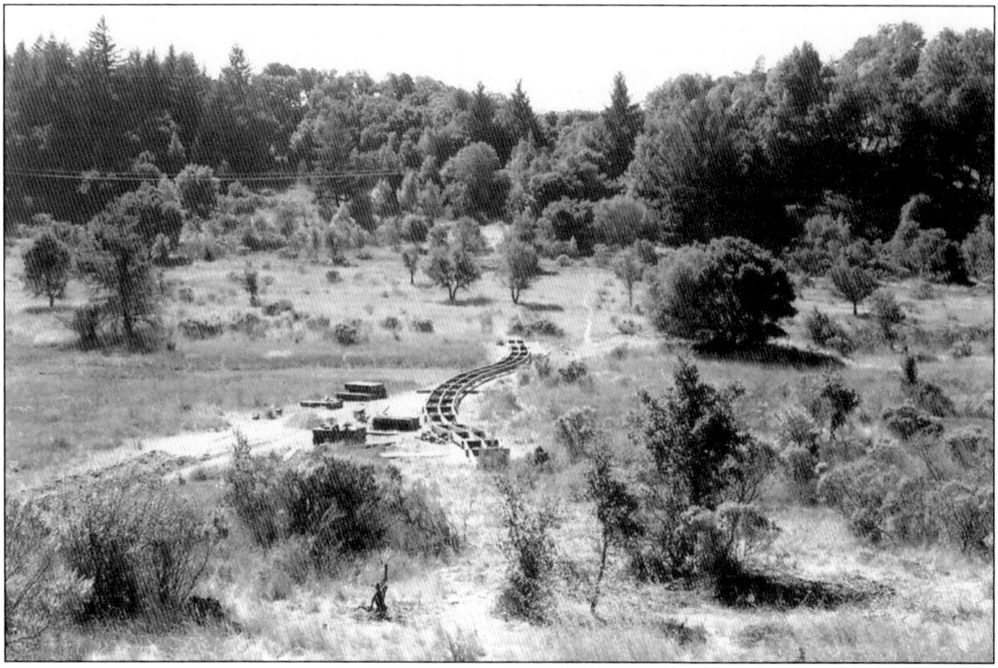

In the 1990s, MMWD started a special fund to accept private donations to help finance watershed enhancement projects such as the boardwalk shown here, both under construction and in its completed state. The boardwalk allows Mount Tamalpais visitors to cross a seasonal wetland area near Sky Oaks Meadow without causing any damage to the fragile native plants located there. In 2011, the MMWD Board of Directors decided to expand the concept of the fund and form a separate, nonprofit, community-based organization. The purpose of the new organization is to provide a way for the community to contribute financial and volunteer resources to support the biological health of watershed lands and to expand water conservation and watershed ecology education programs consistent with district policies.

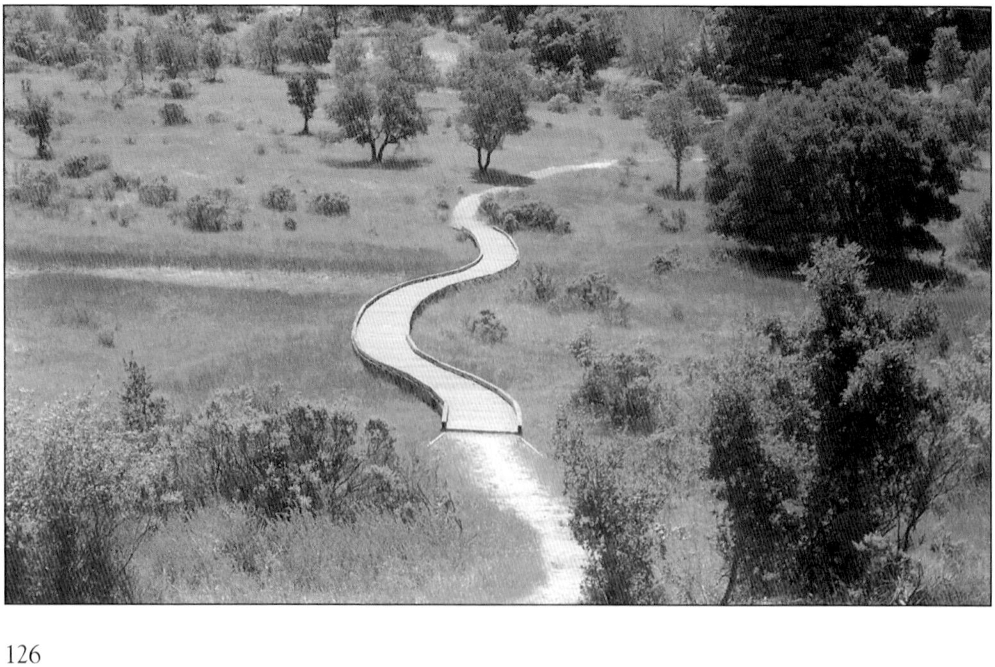

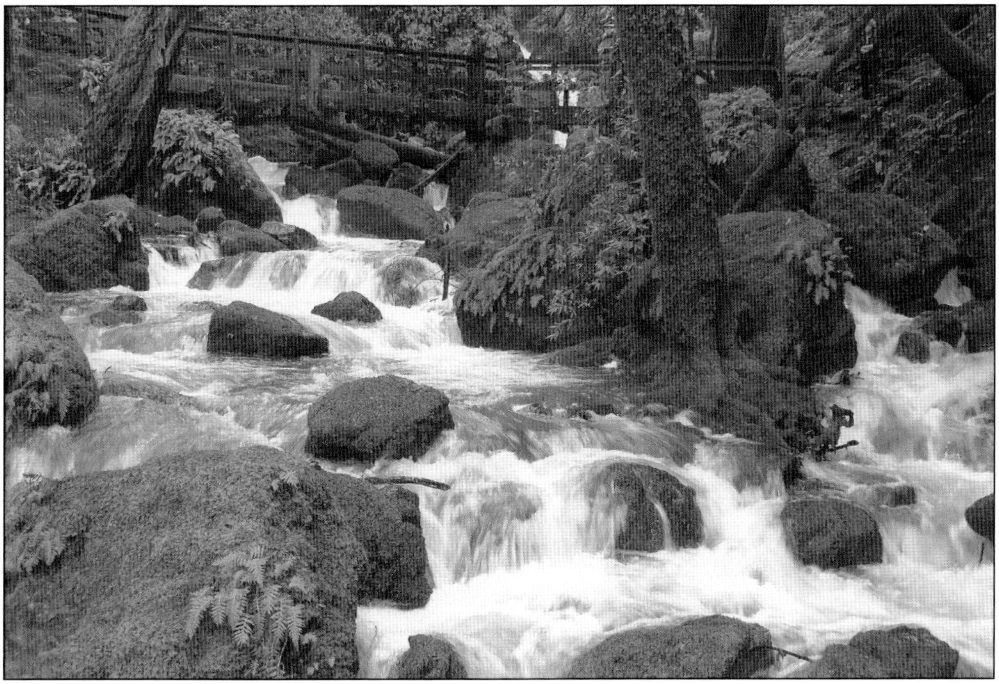

Thanks to the efforts and foresight of MMWD's founders and the continuous stewardship provided by district board members and employees, the Mount Tamalpais watershed is much the same as it was in 1912 and even earlier. Cataract Falls and Trail (above) is located along the north slope of Mount Tamalpais. The trail appears on maps dating as far back as 1898 and was significantly rebuilt with new tread, steps, and bridges by the Civilian Conservation Corps in the mid-to-late-1930s. More recently, it was updated and restored by the Marin Conservation Corps (now Conservation Corps North Bay) under MMWD's direction in the mid-1990s. Bald Hill, below, is one of many distinctive landmarks found on Mount Tamalpais. (Above photograph by Carl Sanders.)

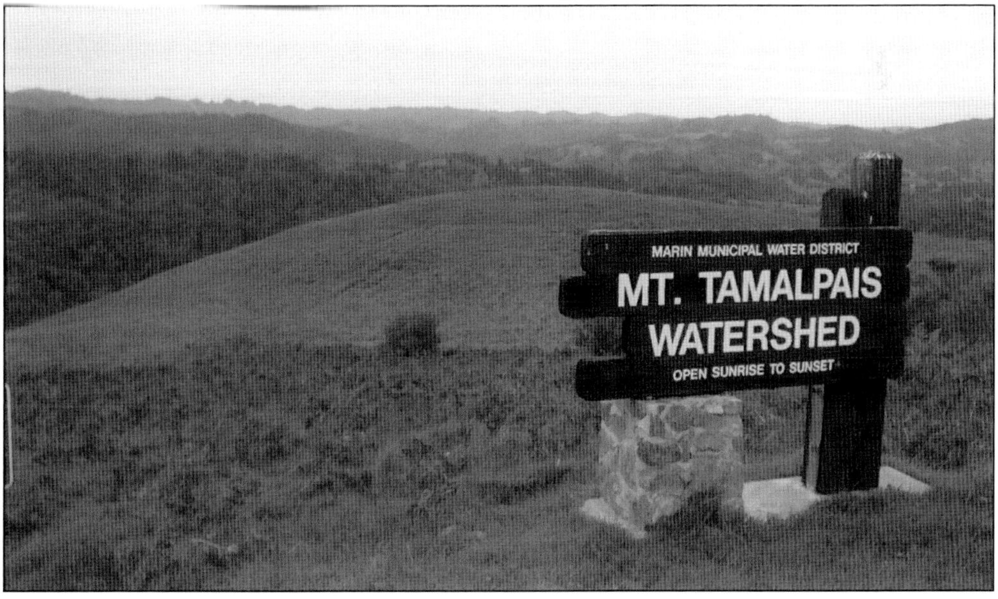